Philosophical Dimensions in Mathematics Education

Mathematics Education Library

VOLUME 42

The titles published in this series are listed at the end of this volume.

Karen François
Jean Paul Van Bendegem
(Editors)

Philosophical Dimensions in Mathematics Education

Springer

Karen François
Free University of Brussels
B-1050 Brussels
Belgium
karen.francois@vub.ac.be

Jean Paul Van Bendegem
Free University of Brussels
B-1050 Brussels
Belgium
jpvbende@vub.ac.be

Series Editor:
Alan Bishop
Monash University
Melboure 3800
Australia
Alan.Bishop@Education.monash.edu.au

Library of Congress Control Number: 2007926609

ISBN-13: 978-0-387-71571-1 e-ISBN-13: 978-0-387-71575-9

Printed on acid-free paper.

9 8 7 6 5 4 3 2 1

springer.com

Contents

Editors and Contributing Authors

EDITORS

Karen François has a Masters Degree in Moral Sciences, a Bachelors of Philosophy from the University of Ghent, and an advanced Masters Degree in Women's Studies from the University of Antwerp. She is currently working at the Free University of Brussels (Vrije Universiteit Brussel) as a scientific researcher for InterUniversity Attraction Pole, Phase V. The project is entitled: "The loyalties of knowledge. The positions and responsibilities of the sciences and of scientists in a democratic and constitutional state." She is preparing her PhD on "Politics of Mathematics." She is particularly interested in the philosophy of mathematics, the politics of mathematics, and mathematical practices in education. In May 2004, she organized an international conference on mathematics and philosophy entitled "Mathematics in Education: Is there Room for a Philosophy of Mathematics in School Practice" at the Free University of Brussels (Vrije Universiteit Brussel).

Jean Paul Van Bendegem is a philosopher, mathematician and professor at the Free University of Brussels (Vrije Universiteit Brussel) in Logic and the Philosophy of Science. He is a visiting professor at the University of Ghent. He is president of the National Centre for Research in Logic, and he is also co-director of the interdisciplinary Centre Leo Apostel at the Free University of Brussels (Vrije Universiteit Brussel). His research concerns mainly non-classical logics, the practice and philosophy of mathematics and, more generally, the relationship between science, religion, and society. Topics studied by members of his own research group, the Centre for Logic and Philosophy of Science, are: cultural identity, the biological origins of language and basic mathematical capacities, the philosophy of information, the philosophy of consciousness, the philosophy of mathematics, practices in mathematics, mathematics education, the politics of mathematics, and the sociology of mathematics.

CONTRIBUTORS

Carmen Batanero is a senior lecturer in the Department of Mathematics Education at the University of Granada, Spain. Before entering this department in 1988, she worked in the Department of Statistics where she taught probability and biostatistics for eleven years. She is presently teaching statistics and statistical education, doing research and supervising doctoral dissertations on statistical education. Her publications include research papers and books directed to teachers. She helped establish the stochastics group at the International Group for the Psychology of Mathematics Education (PME), was co-editor of the Statistics Education Research Journal in 2002-2003, president of the International Association for Statistical Education (IASE) from 2001-2003 and is currently a member of the International Commission for Mathematics Instruction (ICMI) Executive Committee. She was coordinating the International Program Committee for ICOTS-7.

Emeritus Professor **Alan Bishop** was a professor of Education at Monash University between 1992-2002, after spending most of his life in the UK. After completing his PhD in 1969, he was appointed as a lecturer in the Department of Education at Cambridge University where he worked for twenty three years. He edited (1978-90) the international research journal, Educational Studies in Mathematics (published by Kluwer, Dordrecht, Holland), and has been an Advisory Editor since 1990. He is the managing editor of the research book series Mathematics Education Library, also published by Kluwer (1980-present). He is also the chief editor of the International Handbooks of Mathematics Education (1996 and 2002) published by Kluwer. He has been involved with research in mathematics education all his working life, and his Cambridge experience led him to focus attention during the last twenty years on the cultural, social, and political issues involved in mathematics education. He is the author or editor of several influential books, reports, articles, and chapters on mathematics education.

Dimitris Chassapis received his degree in Mathematics from the Aristotle University of Thessaloniki, Greece; his Masters in Science in Mathematics from The City University of London; and his PhD in Educational Sciences from the University of Patras, Greece. After working for seventeen years as a teacher in vocational education and training schools, he was appointed as a lecturer in Mathematics Education in the Primary Education Department of the Aristotle University of Thessaloniki in 1994. He is currently an assistant professor in this department lecturing in undergraduate and postgraduate courses, as well as in continuing education programs for teachers. He has

published several papers on ideological, social, and cultural issues in mathematics learning and teaching in Greek and international journals, and has contributed to international conference proceedings. His current major research interests include socio-cultural aspects of learning and teaching mathematics, as well as epistemological questions on mathematics education as a research field.

Carmen Díaz has a Bachelors of Psychology, a Masters in Social Psychology, and a Diploma in Research Methodology from the University of Granada, Spain. Holding a scholarship from the Spanish Ministry of Education, she is currently working at the University of Granada as a junior researcher for the department of Research Methods in Behavioural Sciences, where she is teaching research methods in Psychology and Education. She is particularly interested in the philosophy of statistical inference, the psychology of probabilistic reasoning, and statistics education. She has published papers on the statistical inference controversy, the assessment of probabilistic reasoning, and applications of statistics in education in Spanish and international journals. She has also contributed to international conference proceedings.

Albrecht Heeffer graduated with a degree in electronic engineering and has a Masters Degree in Philosophy from the University of Ghent. He worked as a researcher on artificial intelligence (AI) at the Turing Institute in Glasgow and at the University of Ghent, Belgium. In 1990, he co-founded a spin-off, specializing in AI and global optimization systems. This company became part of a Dutch-Nordic ICT group in 1999. Since 2000, he has returned to academia to work on the history of mathematics. His doctoral dissertation in 2006 was on the conceptual development of symbolic algebra in the sixteenth century. He published within the domains of AI, mathematics, philosophy, and the history of science. When not engaged in research, Albrecht is an active player of shogi, the Japanese game of chess.

Maria Meletiou-Mavrotheris was born in Nicosia, Cyprus. In 1990, she graduated from the Pedagogical Academy of Cyprus. She then moved to the United States to study at the University of Texas at Austin. She earned the degrees of BA in Mathematics (1993), Masters in Science in Statistics (1994), Masters in Science in Engineering (1998), and a PhD in Mathematics Education (2000). While pursuing her graduate studies, she also worked as a teaching assistant (1993-1995) and a research assistant (1995-2000) in the Mathematics Department of the University of Texas. Upon returning to Cyprus, she worked as a lecturer at the Cyprus Pedagogical Institute (2000-2001), and an advisor for mathematics teachers (2001-2003). She currently

holds the position of assistant professor in the Education Department of Cyprus College. Her research work focuses on the understanding of obstacles to the learning of statistics and on ways of using this understanding to create improved learning environments.

Rik Pinxten is professor of Cultural Anthropology and Comparative Study of Religion at the University of Ghent. He studied philosophy and was a visiting scholar and professor at Northwestern University and the Summer Field School of New Mexico. He has done fieldwork with Navajo Indians, and is editor and co-editor of several journals in the social sciences. His publications include: Anthropology of Space (1983); Evolutionary Epistemology (with Werner Callebaut, 1987); Culturen sterven langzaam (Cultures Die hard, 1994); When the Day Breaks (1997); Goddelijke fantasie (The Creation of God, 2000); De artistieke samenleving (The Artistic Society, 2003); and Culture and Power (with Ghislain Verstraete and Chia Longman, 2004).

Susanne Prediger is professor at the Institute for Research and Development in Mathematics Education at Dortmund University, Germany. She has a PhD in the field of restructuring mathematical logic and her Habilitation thesis is exploring the intersection between philosophy of mathematics and mathematics education, emphasizing the necessity of reflection in the learning process. She has organized several conferences on mathematics and philosophy, has co-chaired the discussion group on philosophy of mathematics education at ICME-10, and is co-editor of the German journal "Praxis der Mathematik in der Schule."

Bart Van Kerkhove holds Masters Degrees in Economics and in Philosophy. He is a research assistant at the Centre for Logic and Philosophy of Science of the Free University of Brussels (Vrije Universiteit Brussel). As a philosopher of mathematics, he is interested in the topic of mathematical practice, more particularly exploring roads to a reconciliation or even (partial) integration of traditional metamathematical and "maverick" sociological accounts of the discipline. In 2002, he co-organized the Brussels conference entitled Perspectives on Mathematical Practices, which had a continuation in the spring of 2007.

Ard Van Moer is a commercial engineer (since 1995) and received a PhD in Philosophy from the Free University of Brussels (Vrije Universiteit Brussel) in 2003. In his doctoral thesis, he studied the intentionality of formal systems, combining the philosophy of mathematics with the philosophy of mind and the mind-body problem. He has been teaching mathematics to

student commercial engineers for almost ten years. His main subjects of interest are the history and philosophy of mathematics and the psychology of the mathematician. At this moment, he is studying the differences and the similarities between mathematics and (conceptual) art.

PRELUDE

A Philosophical Approach to Mathematics

Jean Paul Van Bendegem and Karen François

GETTING THINGS STARTED: AN EXPERIMENT OF THOUGHT

Take a look at the following list:

- mathematics
- education
- philosophy
- society
- history
- sociology
- anthropology

Imagine that this was an IQ-test and a typical question would be: which item definitely does not belong in this list? We expect that the one word at the top of the list, even if it is not the only answer, will be "mathematics". Although there is a connection between mathematics and education and history (after all, there exists a history of mathematics, although this is not proper mathematics), the links between the six other members of the list are certainly more recognized than that between each one of them and mathematics.

Let us do a mental exercise, however, and look at all the various combinations that include mathematics. Because each of the remaining six items on the list can be examined in relation to mathematics, this gives us 2^6

(i.e., 64) possibilities. One might argue, however, that society and history-sociology-anthropology can be grouped into a single larger unit, called "society-plus", in the way that we have society and its reflective study both on the level of individuals and in groups in a temporal-historical and cross-cultural perspective. So, that leaves us with:

- mathematics
- education
- philosophy
- society-plus

It is tempting to further develop this line of thinking. We could imagine all the pairs that can be made out of these four, with mathematics as one of the elements, then look at all the triplets we can form and finally the quartet itself. We will resist this move, however, for the simple reason that the collection of papers brought together here have the interesting quality that at least three of the four (if not all four) aspects are somehow covered. It would, therefore, be quite arbitrary to *invent* a framework so each one of the contributions can be inserted into a slot, making it necessary to misrepresent (more or less) its contents. Perhaps it is a rather bold claim, but we do believe that mathematics, education, philosophy, and society-plus, (*MEPS* for short), form a connected whole and should preferably be approached as such.

In addition, we would like to emphasize the presence of the letter *P* in *MEPS*. Mathematics, education and society are well-studied, but we have the impression that the case is not entirely the same for mathematics, education and philosophy. As to the former, it does not—or more strongly, should not—come as a surprise that society-plus enters into mathematics through education. After all, an educational system, whatever particular shape it takes, involves society at large and involves ways to transmit knowledge, practices, and social skills from one generation to the next. It is no small wonder then that we should reflect on how mathematics is imbedded in society. As for the latter, the relationship between the three elements is not at all clear. It ranges from a trivial relationship such as a simple philosophy of mathematics (a philosophy of education is always possible if not already existing) to a relationship that has received little attention up to now (e.g., can philosophy be integrated into a mathematics curriculum for 12-year olds?). Philosophers themselves are not truly impressed by educational problems, regardless of the area we are talking about. Tongue in cheek, it seems that philosophical knowledge has a curious property in common with mathematical knowledge. It seems to be passed on from one generation to the next directly through *sublime* or *fine* adult minds without the usual

detour of a long, cumbersome education. One is tempted to think that eternal truths (as so many among us wish to see mathematical knowledge consist of) will somehow seek out the *fine* and *exquisite* mind and give it direct access to the ideal mathematical realm without requiring any ediucation.

Let us elaborate on this thought a bit further. These questions have been asked a million times before. Nevertheless, let us ask again: Why do we consider mathematical knowledge and possibly mathematical practices so important that generation after generation must undergo the extreme displeasure of calculating fractions, square roots, derivatives, or integrals? Remember the famous quote "When was the last time *you* had to calculate the determinant of a matrix?" This assumes that one is still capable of performing this small miracle of numeracy. Indeed, should these capabilities not be reserved for those who have a privileged access to them—i.e., the mathematically *gifted*? Or, in other words, should mathematics education not promote selection?

And, for that matter, should we be fighting segregation at all? Is selection not inevitable in education? After all, there will always be bright pupils and weak pupils. Should we not then apply a method that allows the brightest to surface as quickly as possible, so as to be able to offer them the best education and reduce the suffering of the not so bright? Until a few decades ago, segregation was still produced on the basis of an elite language. For example, in some European countries, Latin was the language of knowledge and was by no means accessible to all because it was not the language of the people. Has mathematics not taken over this role in many countries as a kind of *lingua franca* for the sciences? The differences between bright pupils and poor pupils will remain, but that does not mean that we must endeavor to generate this segregation as quickly and as sharply as possible. As an alternative, would it not be better to make our ideological-philosophical project known and to indicate that we are advocates—or at least proponents—of an emancipatory education that helps to develop democratic values, via mathematics education, as well? Just as people have a right to literacy, do they not have a right to numeracy as well? These questions are more frequently raised now as internationally promoted measurements of the prosperity of a population include measurements of the mathematical capacities of that population. This does not imply that differences between strong and weak pupils are ignored. After all, everyone is helped by the best education and, in fact, is even entitled to it. The point is that mathematics education in this project should not be a means to strengthen, let alone to justify, the current divide.

In summary, as Alan Bishop and others have so clearly shown, a set of values, a view of society, an implicit philosophy—basically a way of looking, perceiving, being in the world—are also transmitted at the same

time. Whatever one's perspective or view on this matter, gaining insight into this conceptual cluster is of the utmost importance.

THE CORE THEME OF THE BOOK: MAKING THE IMPLICIT EXPLICIT AND ASKING QUESTIONS

Given such a set of values, two questions come quite naturally: 1) whether these values, if implicitly present, can be made explicit, and 2) if such values are found wanting, whether alternative values exist and, if so, how they could be implemented? All contributors to this volume share this thought—that the mathematics being taught in Western culture carries with it a *questionable* philosophy and world-view and that alternatives do indeed exist.

The contributions presented in this volume have more is common. What we understand by "philosophy of mathematics" is not restricted to the meaning of philosophy in the sense of axiology. When speaking about implicit philosophy, we strongly emphasize the aspect of values. While speaking about explicit philosophy, we are searching for ellaborations on the notion of philosophy. What all these articles share in addition is the strong belief that critical philosophical reflections *should be included* in the standard mathematics curriculum. At the same time, it should be noted that it is less clear what particular form this implementation should take.

First, are we thinking about philosophical systems, theories, views, and the like, or do we see philosophy as (a set of) practice(s)? Do we see pupils as *little philosophers* or rather as human beings with a philosophical attitude? Related to this point, are we looking for a philosophical system that supports, explains, and motivates the mathematics taught—or, quite the opposite, do we wish to confront the pupils with critical remarks and observations about the mathematics they have learned?

Second, are we striving for an explicit presence of a philosophical reflection in the curriculum, or does the philosophical content enter in somewhat indirect ways? Using the history of mathematics—or is *a* history of mathematics a more apt expression?—to show the flexibility of mathematical concepts is, after all, a nice way to induce a philosophical-critical attitude concerning mathematics.

Third, it seems obvious to us that all parties concerned with the educational process should be involved in the process. We tend to focus on the curriculum and on the pupils, but there is really no reason to exclude the teachers themselves. Ideally, one could imagine a philosophically motivated mathematics teacher who, given a classic curriculum, manages to present the

material in such a way that the pupils *sense* or *feel* the teacher's philosophical attitude.

All these considerations lead to the following connected set of questions:

- What is the philosophical set of values implicitly present in mathematics curricula today?
- Conversely, what are the views on mathematics education from the viewpoint of present-day philosophies of mathematics (Note: it is our belief that this question is hardly ever asked)?
- Given answers to the above questions, which philosophies are compatible with which kinds of education, and what specific form can such philosophies take?
- How can such philosophies be explicitly included in the curriculum? What particular shape can such implementation take?
- If all of the above can be, one way or another, satisfactorily answered, how are we to evaluate afterwards?

In short, this sequence of questions corresponds to a movement away from an implicit philosophy *of* mathematics towards an explicit philosophy *in* mathematics. Each in his or her own specific fashion, the authors have tackled one or more of these questions, either by focusing on broad theoretical reflections or by presenting quite concrete ideas for actual, alternative practices—in some cases, those already implemented. What follows is a detailed introduction to the individual papers in this volume, interspersed with the editors' philosophical and other commentaries.

OVERVIEW OF THE PAPERS

The paper by Karen François, "*The Untouchable and Frightening Status of Mathematics: Didactics, Hidden Values and the Role of Ethnomathematics in Mathematics Education,*" explores the connection between the absence of an explicit philosophy and the implicit philosophy of a curriculum. In the case where an explicit philosophy is mainly absent, the implicit philosophy seems to be rather absolutist with no room for reflection or critical attitudes. She criticises a curriculum which presents mathematics as a series of merely technical procedures without an embeddingof culture.

The papers by Susanne Prediger, "*Philosophical Reflections in Mathematics Classrooms. Chances and Reasons;*" Dimitris Chassapis, "*Integrating the Philosophy of Mathematics in Teacher Training Courses. The Relevant Problems Encountered in Greece as an Example;*" and Albrecht Heeffer, "*Learning Concepts through the History of Mathematics:*

The Case of Symbolic Algebra" contain several proposals, ideas, and suggestions as to how to steer the mathematical education process in other directions. What will strike the reader most is the variety of possibilities. As Prediger suggests, you can introduce philosophy directly into the classroom. She explains why and how philosophical reflections should be included in mathematics classrooms by presenting three examples. Chassapis would like to see more emphasis at the teachertraining level and has actually put this into practice. The philosophy of mathematics should be considered an essential component of teachers' professional knowledge. He presents some examples of courses in learning and teaching primary school mathematics, and issues that arise in its implementation. Heeffer has a dream at the level of the history of mathematics (and note that it is an historical tale with philosophical consequences, for he questions the invariance of mathematical meanings of concepts over the ages). He argues for the integration of the history of mathematics within the mathematics curriculum as a way to teach students about the evolution and context-dependency of human knowledge. Moreover, an emphasis on the understanding of mathematical concepts is a necessary condition for a philosophical discourse about mathematics.

The above characterization wrongly suggests that to solve the problem at hand it is basically sufficient to introduce some additional topics, such as a philosophical reflection or an historical tale. One could (and should) also criticize the way it is presented—how mathematics is carried out. An obvious is to emphasize the problem-solving nature of any intellectual activity, hence also mathematics. What is more central to problem-solving than learning from errors? This topic is dealt with in a number of papers (e.g., Chassapis, Meletiou-Mavrotheris, and Prediger), but Carmen Batanero and Carmen Díaz's paper, "*The Meaning and Understanding of Mathematics: The Case of Probability,"* is a truly fine example. They distinguish between the personal and the institutional meaning of mathematical concepts to differentiate between the meaning that has been proposed for a given concept, or fixed in a specific institution, and the meaning given to the concept by a particular person in the institution. Furthermore, they describe mathematical activity as a chain of semiotic functions and introduce the idea of semiotic conflict that can be used to give an alternative explanation to some widespread probabilistic misconceptions. The paper also contains an excellent critique of the fact that the probability theory is often presented as if there exists one and only one interpretation, whereas a first-order philosophical exploration reveals at least five or six fundamentally different meanings.

There is, however, no need to stop here. It is a fruitful and (we believe) rather original exercise to take the philosophy of mathematics as a starting point and, having a particular view on what mathematics is and how it is practised, ask what implications it has for an educational view on

mathematics. If one is a Platonist or a formalist, then it is to be expected that there will be a focus on mathematical results, such as theorems and proofs. If an historical component is included, it will tend to focus on the great achievements of great mathematical, and almost exclusively masculine *fine* and *exquisite* minds (one may safely ignore the mathematicians' bodies). In these last few decades, there has been an emphasis—and this is a rather novel development—on mathematical practice and thinking through its philosophical implications.

In this collection of papers, the contributions by Maria Meletiou-Mavrotheris, "*The Formalist Mathematical Tradition as an Obstacle to Stochastical Reasoning,*" Ard Van Moer's "*Logic and Intuition in Mathematics and Mathematical Education,*" and Bart Van Kerkhove's "*A Place for Education in the Contemporary Philosophy of Mathematics: The Case of Quasi-Empirism*" explore this dimension. Van Moer draws attention to the faculty of mathematical intuition not as a gift of the gods or a supernatural ability or, to follow today's fashions, genetically given, but as the result of a learning process—hence something that can be thought. Van Kerkhove shows that at the basis of Lakatos's quasi-empiricist philosophical excavations lies a central concern, through his affinities with Pólya, with pedagogical principles surrounding the informal phases of mathematical inquiry, which suggests that the development of humanistic mathematics is indeed not unconnected with the educational. Meletiou-Mavrotheris criticises the continuing impact of the formalist mathematical tradition. She reconsiders some empirical findings on students' understanding of statistics, and forms some hypotheses regarding the link between student difficulties and mathematical formalism. Subsequently, she suggests the pedagogical and curricular changes that ought to take place in order to move away from the formalist mathematical tradition and to improve people's impoverished probabilistic and statistical reasoning.

Have we now reached a point where our explorations can temporarily end, or are there further dimensions to study, scrutinize and criticize? In fact, we believe there are. Better still, we believe we have to. Many of the papers in this book deal with it implicitly, but the paper by Rik Pinxten and Karen François, "*Ethnomathematics in Practice*" does so explicitly (thus making a full circle). It tackles the question of the relationship between mathematics with the big *M* or eurocentric mathematics, and ethnomathematics with the small *m* or usually non-Western systems of mathematical thinking and doing. There is a major difference to note here. As transpires from the above formulation, *our* mathematics comes with a capital *M* and all non-Western mathematics comes with a small *m*.

In line with all the contributions already mentioned, it is a far more interesting hypothesis to explore that big *M* and small *m* are also present in

our Western world and are thus translated into mathematical education. If one explicitly considers mathematical thinking as the highest, or the *purest*, or the *noblest*, or the *most abstract* form of human intelligence achievements, then it seems inevitable that he (sometimes she) who masters this superior form of knowledge somehow already had the disposition for it. In short, as Alan Bishop has shown, there is a difference between *m*(athematics) and *M*(athematics), where *m*(athematics) stands for a set of mathematical basic competence (such as counting, designing, explaining, locating, measuring, and playing), and *M*(athematics) stands for mathematics as it is known and developed as a Western scientific discipline. It is the mathematical knowledge of the mathematician and of the highly specialised student in physics or engineering. In contrast to this, *m*(athematics) is a set of mathematical basic skills and procedures that an individual or group uses in daily life. It is a clear invitation to connect the dividing line between the mathematically initiated (the "happy few") and the rest of the world—i.e., most of us with the "mathematics for the million" (although the splendid work by carrying this title, is really a scaled-down version of what the happy few think—a form of *Theology Made Easy* or *Theology for Dummies*) —to social and political dividing lines. We did not make a mistake here. We did write *theology* and not *mathematics*. This introduction is not the right place to pursue these matters, but we dare to provoke the thesis that the similarities are more striking than the differences.

Another route not to be explored here is that the big *M* in an educational setting has a curious and complex relationship what actually happens in the mathematical research community. Either mathematicians are idealized, losing all connection with blood-and-flesh mathematicians, or the focus is mainly on the *fossilized* results—again breaking the link with the messy (and usually more interesting) practice of mathematicians. This strongly suggests that a distorted *M* is presented to the pupils.

The example of Navajo geometry in the contribution of Pinxten and François serves not so much to illustrate that other mathematical systems (to be understood in the broadest terms possible) are possible—we consider this question to have been dealt with, and, if one does object, we believe that a definitional game is played along the lines of "I don't care what it is they do, as long as you do not call it mathematics"—to throw light on the Western cultural situation. The example of the Turkish ethnic minority shows how ethnomathematics comes into the picture in Western classrooms. After all, Western societies have transformed into multicultural societies, hence ethnic minorities are part and parcel of everyday life. Where better does one find its reflection than in an educational setting? The issue at stake is how to deal with diversity in the classrooms and more specifically in the mathematics classroom. We have to be aware of the institutional context of mathematics

education. The point seems to be that mathematical learning or thinking is contextual in any living culture; it lives and develops and it is used in a particular cultural context. The often decontextualized use of knowledge in a Western school setting is then alienating and foreign to pupils. And so the *S* in *MEPS* again enters into the picture, demonstrating what we claimed at the start—that *ME* without *S* does not really make sense and that *ME* without *P* does not make much sense either. This last statement should be taken seriously in our minds (and bodies for that matter), for as we hope this volume shows, thinking about *P* in the context of *ME* leads one straight back to *S*, thus making the impact of this component clearer and more pertinent.

INTERLUDE 1

When we are talking about a philosophy of mathematics and implementing it in general in a mathematics curriculum, it is good to first ask how mathematics curricula are doing. Every one of us has a vague idea about what mathematics education should look like. Perhaps our memories go back to our own training where only a rara avis had another mathematics education rather than a technically oriented training. The history of mathematics was probably limited to an interlude in a handbook showing a gallery of male mathematicians, that we were not expected to memorize and who we crossed out, marking "don't study." What has become of the current mathematics curricula? In the following article, we read about a study of the mathematics curriculum in Flemish secondary education (Belgium). The Flemish mathematics education outshines that of most countries in international rankings—certainly for Trends in International Mathematics and Science Study (TIMSS), but for Programme for International Student Assessment (PISA) too. In this respect, the Flemish mathematics education could be a model for other countries. But alongside the success story of Flemish mathematics education there is also a—perhaps not uninteresting—critical story to be told. This critical story is a perfect introduction to a philosophy of mathematics in education.

This article does not just concentrate on the explicit philosophical aspects which may, or may not, be part of the curriculum. The absence of an explicit philosophy, too, is a story in itself. Despite the absence of explicit philosophical aspects, a philosophy indeed hides behind the curriculum. In fact, a curriculum is the result of a lengthy debate among experts about whether or not to include certain aspects. There is an underlying view that suggests some interpretations about the perspective on mathematics with regard to society. The question remains whether the perspectives of the experts responsible for developing the curriculum also filter through to what pupils are eventually expected to know in class. After all, the curriculum excels in creating an accumulation of technical mathematical facts. With an ever-increasing complexity, the various subdisciplines of mathematics are pieced together and systematically taught, practised, practised and... practised. Their relevance and the links to the pupils' social world are generally absent. Yet, this approach is characterised by its underlying values, and it is important to uncover these—not just to dispel the idea that mathematics is something neutral and objective, with no link to human interests, but to show how these values remain reserved for a select group, through the way that they are transferred, of elite pupils who understand abstract mathematics without need of concrete applications.

THE UNTOUCHABLE AND FRIGHTENING STATUS OF MATHEMATICS

Didactics, Hidden Values, and the Role of Ethnomathematics in Mathematics Education

Karen François[1]
Centre for Logic and Philosophy of Science – Free University of Brussels, Belgium

Abstract: During my research into the mathematics curriculum of Flanders secondary education (age 12-18), I first discovered that there is small scope for an explicit philosophy of mathematics. Nevertheless, there are some initial concepts formulated in the general objectives which tend to a more absolutist view of mathematics. In formulating the new curriculum, however, there was some attention paid to the inclusion of humanistic values. The mainstream of the implicit philosophy of mathematics is still a rather absolutist one, viewing mathematical truth as absolute and certain, connected with some humanistic values. Second, I discovered a large gap between general and vocational education. On the one hand, we can say that mathematics in vocational education is completely embedded in a modular system, and that attention is paid to core skills. On the other hand, we must say that pupils are prepared for specific occupations, for personal and social functioning, and to survive in our society. Access to higher education is theoretically possible, but unlikely for the majority. Mathematics in general education is a separate and different course. General education provides a strong base for higher education. In this paper, I shall briefly present some findings of the case study in which I want to make the connection with the theoretical framework of Alan J. Bishop. Bishop believes in a difference between the small *m* and the large *M* of mathematics, where the small *m* stands for a set of mathematical basic competence (such as counting, designing, explaining, locating, measuring, and playing) and the large *M* stands for mathematics, as in the Western scientific discipline. I shall argue that pupils in vocational mathematics are taught the small *m* and pupils

[1] Research for this paper was done within the framework of the Inter University Attraction Poles program (Phase V) granted by the Belgian Federal Science Policy.

in general education are taught the large *M*. The more general the education is, the larger the *M* is, which gains higher respect in society. In line with this *M-m* distinction, I shall elaborate on the connection which exists between the view on mathematics education and the didactics used in classroom. International comparative research on the results of mathematics educations shows us—in the case of Flanders—the best results nearly all over the world. I shall, however, criticize the way in which mathematics in schools chooses between the smart and the not so smart and what the role of ethnomathematics should be in Western school curricula to overcome this social stratification. Finally, I want to go on to explore two central hidden values in mathematics education to demystify the untouchable and frightening status of mathematics.

Key words: Mathematics education, philosophy of mathematics, hidden values, ethnomathematics

1. INTRODUCTION

Research on the implicit and explicit philosophy of mathematics in school curricula takes place in a broader research project, in which we are looking for the relationship between the sciences, society, politics, and the democratic constitutional state. Within this project, one of the key questions are: "What is the place of mathematics in (or indeed above?) the sciences" and more broadly, "What is the place of mathematics in society?" By narrowing this question down, we shall elaborate on the question of how mathematical knowledge is reproduced in our society and how mathematics is handed down from generation to generation. Obviously, education is an important, if not the most important way to reproduce knowledge in our society.

In a first stage of the research, I was looking for the answer to "is there room for a philosophy of mathematics in school curriculum?" At a later point, this question seems rather simplistic. So does the question "Why should we implement a philosophy of mathematics in the mathematics curriculum?" because there already exists a philosophy of mathematics in the current curriculum. It has not yet been made explicit, however—it remains hidden. Moreover traditional mathematics is strongly directed towards the performance of techniques and has little to do with the study of mathematics as a historical and cultural product nor with the underlying cultural values. If we make the difference between an implicit and an explicit philosophy, we can propose the question, "Is there room for an explicit philosophy of mathematics?" The existence of an implicit philosophy is obvious, as I shall argue in the last section on hidden values in mathematics education. The absence of an explicit philosophy of mathematics tells us

something about the implicit values of the curriculum. To discover them, we need to implement an explicit philosophy. If an explicit philosophy is generally absent, the implicit philosophy will be a rather absolutist one with no room for reflection or critical attitudes. I will show this statement in the following chapters, where I shall first present some main findings on the screening of the mathematics curricula. Second, we will search for a deeper understanding of these results by presenting three main aspects of the curriculum. Finally, I shall criticize the current curriculum that presents mathematics as a series of technical procedures without a cultural embedding. Our research shall be based on the study of relevant literature on philosophy, mathematics and on education (Bishop [1988a] 1997, Bishop (2002), Bishop (2006) and Ernest [1991] 2003).

2. CURRICULUM SCREENING

During my research into the curriculum of mathematics in Flanders secondary education, I discovered three main aspects: 1) the absolutist view on mathematics, 2) the gap between vocational and general education, and 3) the related distinction between mathematics with a large *M* and mathematics with a small *m*. Before arguing these aspects, we will give an overview of the retained philosophical issues.

The method used to analyze the curriculum was a qualitative screening at two levels:

- the level of the curriculum as developed by the community (as strictly enforced by law)
- the level of the authorities of the differing school systems

In Flanders, there are three differing school systems—public schools, subsidized private schools, and subsidized community schools. They have to integrate the attainment targets into their own developed curricula.

Before we present the kinds of philosophical items in the curriculum, we need to point out that there are two parts of the curriculum where philosophical issues can be found. The first part is the view on mathematics in education and some general objectives. The second part is the attainment targets. It is understandable that teachers are focused on Part Two because the attainment targets are the criteria for the evaluation of pupils.

Secondary education has four forms: general, technical, art, and vocational. The four forms of education are not organized separately in the first stage (12-14-year olds), but they are in the second stage (14-16-year olds) and in the third stage (16-18-year olds). In grade I, there is an A class

that has access to the general, technical and art forms and there is a B class that only has access to vocational education classes.

The research question is whether there are targets—beside the technically oriented targets—integrated in the curricula concerning philosophical issues in the broad meaning of the term (e.g., (strictly) philosophical, cultural, historical, and so on).

In the following tables, you will find listings of retained paragraphs on philosophy (in the broad meaning of the term) and the non-technically oriented targets.

2.1 The Curriculum as Developed by the Community at the Level of the View

In a first general overview (Table 1), one can see that there is no room for philosophy in vocational education. As we shall explain in section 3.2, there is a real gap between general and vocational education.

Table 1: "Philosophy"[2] of Mathematics at the Level of the View. An Overview

Grade	**Type of Education** Types of education are shaded where philosophy occurs.			
I	**A-Type General**			**B-Type Vocational**
II	**General**	**Technical**	**Art**	**Vocational**
III	**General**	**Technical**	**Art**	**Vocational**

Now we will go into the philosophical issues in more detail. Table 2 presents a listing of all issues referring to non-technical goals in the curriculum. We differentiate by grade (I, II and III, that correspond

2 We placed *philosophy* between apostrophes because we needed to use a very broad interpretation of philosophy.

respectively with the ages of 12-14, 14-16, 16-18) and by type of education (general, technical, and art), if necessary.

Table 2: "Philosophy" of Mathematics at the Level of the View. Detail

Grade I: A-Type (General Education)
Ontological proposition: The proposition that mathematics is abstract and formal and that mathematics has no connection with reality, has increased to a certain degree. **Appreciation**: Pupils must be encouraged to see the beauty and perfection of a geometric figure, the clarity of a well-reasoned argument, and the elegance of a formula. **The cultural and dynamic meanings of mathematics:** Pupils should experience that mathematics has a practical use, and that it has educative and aesthetic value. The history of mathematics helps pupils to understand that mathematics is an important aspect and component of culture, both in the past and the present. Mathematics in the past developed via many cultures. Due to the emphasis of this development, pupils gain the knowledge that mathematics is a dynamic process. **The fundamental goals are:** Pupils will have the experience of mathematics as a dynamic science. Pupils will have the experience of mathematics as an important cultural component.
Grade II: general, technical and art education
The ontological proposition: Absent. **Appreciation:** In addition, when the commission determined the selection of the goals, they took the effect of the development of a relationship with mathematics into account.

The cultural and dynamic meanings of mathematics: (more abstract): Pupils should experience that mathematics has a practical use, and that it has educative and aesthetic value. Attention to the development of mathematics helps pupils understand that mathematics is an important aspect and component of culture, both in the past and the present. In this manner pupils will gain the knowledge that mathematics is a dynamic process.

The fundamental goals are: Pupils will have the experience of mathematics as a dynamic science. Pupils will have the experience of mathematics as an important cultural component.

Grade III: General Education

The same content as in Grade II.

Grade III: Technical and Art Education

Text [*between square brackets*] is dropped at this level.

The same content as in Grade II.

The cultural and dynamic meanings of mathematics (a partial interpretation): Pupils should experience that mathematics has a practical use, and that it has educative and aesthetic value. [*Attention to the development of mathematics helps pupils understand that mathematics is an important aspect and component of culture, both in the past and the present. In this manner pupils will gain the knowledge that mathematics is a dynamic process.*]

The fundamental goals are (one goal has been dropped): Pupils will have the experience of mathematics as a dynamic science. [*Pupils will have the experience of mathematics as an important cultural component.*]

In section 2.2, we move on to the more important level of the attainment targets. While it is possible to ignore the viewpoint and general objectives of a curriculum, teachers are supposed to take into account the attainment

targets. They have to focus on that part, because the attainment targets are the criteria for the evaluation of pupils and textbooks are created based on these targets.

2.2 The Curriculum as Developed by the Community to Reach the Attainment Targets

Looking at the overview of possible locations for philosophical issues (see Table 3), one will see there are no philosophical (historical or cultural) goals formulated for either the B type (vocational education), or for grade I. The philosophical issues are reserved only for grades II and III (see Table 3) of general education.

Table 3: "Philosophy" of Mathematics at the Level of the Attainment Targets. An Overview

Grade	**Type of Education** Types of education are shaded where philosophy occurs.			
I	**A-Type General**			**B-Type Vocational**
II	**General**	**Technical**	**Art**	**Vocational**
III	**General**	**Technical**	**Art**	**Vocational**

Table 4 presents the listing of the non-technical attainment targets of the curriculum. We only retained three different attainment targets over the six years of secondary education. Philosophical goals are completely skipped in vocational education and in Grade I (age 12-14). While there is some attention paid to a philosophy of mathematics at the level introductory of the curriculum, one can see that at the level of the content of the course—the real stuff that pupils have to gain—there is very little room for philosophical reflection.

Table 4: "Philosophy" of Mathematics at the Level of the Set Attainment Targets. Detail

Grade II: General Education
Pupils can give examples of the contribution of mathematics to art.[3]
Grade II: Technical and Art Education
Pupils will gain an appreciation for mathematics (possibilities and limitations) in confrontation with the cultural, historical, and scientific aspects of mathematics.[4]
Grade III: General Education
The same content as in Grade II: General Education
Grade III: Technical and Art Education
Pupils can give examples of the application of mathematics in other courses and society in general.[5]

3 Attainment Target 8 of the General attainment targets.
4 Attainment Target 11 of the General attainment targets.
5 Attainment Target 7 of the General attainment targets.

2.3 The Curriculum as Developed by the Different School Organizations

It is interesting to see what happens at the level of the self-developed curriculum of the different school organizations. They have the freedom to add some targets, to fill in the content of the attainment targets, and of attitude. What did they do with this freedom? We shall not repeat those aspects that are integral to the self-developed curricula. We shall only pay attention to the new aspects that have been added.

First, it can be said that they (unfortunately) do not appear to have done a great deal with this freedom. Only the curriculum of the Catholic schools displays an explicit ideological message that is directed at teachers of mathematics:

"A teacher of mathematics at a catholic school will teach the same mathematics as their colleagues in public and community schools. However, they have a duty to refer to the ideological project, wherever they can. As a member of the Christian pedagogical project, they should be alert in order to seize any opportunity to emphasize a deeper and more intense dimension. Also the course in mathematics opens up other possibilities. The better teachers know their pupils personally, the better they can feel the moment when pupils have the openness to advance toward ontological and existential questions." (Licap 1997, 14, my translation).

Looking at the level of the attainment targets, we discovered the following three additional philosophical issues:

Grade II (Public School): General, Technical and Art Education
Mathematics can contribute to expressive-creative education, specifically to architecture and the art of painting and sculpture.

Grade II (Catholic School): General, Technical and Art Education
The education of mathematics is bound with other disciplines and courses. Moreover, even mathematics has been developed in a historical context with its specific ideas and problems during the past centuries. It is also important, therefore, to pay attention to the historical context to assist the pupils in gaining an understanding of mathematical problems.

Grade II (Catholic School): Only for General Education
In geography for example, a teacher can pay attention to the contribution of mathematics in architecture, music, painting and sculpture. (Escher, Vasarély, Mondriaan, Roelofs, Le Corbusier, the Pantheon …)

3. FINDINGS OF THE CURRICULUM SCREENING

Based on the theoretical framework of Bishop and Ernest, (Bishop [1988a] 1997, Bishop (2002), Bishop (2006), and Ernest [1991] 2003), we will argue why we retained the following three aspects that characterize the curriculum: 1) the absolutist view on mathematics, 2) the gap between vocational and general education and 3) the related distinction between mathematics with a large *M* and with a small *m*.

3.1 The Absolutist View on Mathematics

By screening all the curricula, we discovered that the more specific the targets were, and the more important the impact on the educational process was, the fewer philosophical issues we found. There is a limited scope for an explicit philosophy of mathematics, especially if we are looking at the level of the specific objectives. Nevertheless, there are some initial concepts formulated at the level of the view on mathematics in education that give us an argument for supporting the existence of some humanistic values.

The initial concepts formulated at the level of the view, however, are not translated to attainment targets. This means that the mainstream of the current implicit philosophy of mathematics is still a rather absolutist one—considering mathematical truth as absolute and certain (See Pinxten and François, this volume).

The absolutist view on mathematics is defined by Ernest as "it consists of certain and unchallengeable truths. According to this view, mathematical knowledge is made up of absolute truths, and represents the unique realm of certain knowledge, apart form logic and statements true by virtue of meanings of terms, such as 'All bachelors are unmarried'" (Ernest 1991, 7).

The current curricula do not pay attention to reflection, critical attitudes, or the social construction or cultural variety of mathematics. Mathematics is handed down as if there exists only one mathematics that is made up of absolute truths and certain knowledge where specific deductive methods provide the warrant for the assertion of mathematical knowledge.

In support of the absolutist view, offer the following arguments:

- There is no room to discuss the status of mathematics
- The status is very clear and rather static
- There is no philosophy at all in vocational education
- The larger the *M*, the higher the respect in society
- The appreciation for mathematics that pupils are encouraged to gain is seen as the highest form of motivation

- Experience-based learning is only used to gain the interest and to motivate disinterested pupils, to help them to gain appreciation for mathematics with the truly large *M*

As for the humanistic values, we observed the following:

- There is only a small space for philosophy in education in general
- There is limited attention given to "the possibilities and the limitations of mathematics," although in the curriculum it is placed between brackets
- Some attention is given to the applications of mathematics
- There is limited attention to historical and cultural components (where, in addition, most of the space is filled with art)

3.2 The Gap between General and Vocational Education

As presented previously (Tables 1 and 3), there is not even one reference to an explicit philosophy in the curriculum of vocational education. That creates a large gap between general (technical and art) and vocational education. Pupils in vocational education are prepared for specific occupations, for personal and social functioning, and to survive in our society. Therefore, it seems to be that they do not need any critical or philosophical reflection on mathematics, nor on the history or the cultural embedding of it. Access to higher education is theoretically possible, but unlikely for the majority, while general education provides a strong base for higher education. Hence, vocational and general education have a truly differing status and this brings us to our third main point.

3.3 The More General Education is, the Larger the M Value is

Alan J. Bishop distinguishes between the small *m* and the large *M* of mathematics, where the large *M* stands for mathematics as the *Western* scientific discipline, and the small *m* stands for *universal* mathematical basic competencies. What does Bishop mean by the small *m*? Bishop ([1988a] 1997, 20-60) presents six key universal activities that are the foundations for the development of mathematics in culture—namely counting, locating, measuring, designing, playing, and explaining. He claims that those six activities are universal and therefore mathematics is a pan-cultural phenomenon. With the help of the available cross-cultural evidence, he explores the hypothesis that these six activities are *universal*, and yet one cannot establish whether such activities are indeed universal. One can only

infer from the available evidence. Hence, *plausibility* is a reasonable criterion to use here so that one can argue that it is at least plausible that the six activities are universal. Furthermore, Bishop demonstrates that all cultures have necessarily developed their own symbolic technology of mathematics in response to the demands of the environment. In other words, mathematics is a culturally embedded human activity. It is not only *universal*—it is at the same time a pan-cultural phenomenon. Bishop argues that the six key activities are, and were, the significant activities for the development of mathematical aspects of culture. As a result of certain within-cultural developments, and also of different cultures interacting and conflicting, a particular and traceable line of development has emerged. An interesting example of the evolution of the concept of number is given by Heeffer (see Heeffer, this volume).

In terms of Bishop's distinction between the small *m* and the large *M*, pupils in vocational education are taught a form of mathematics that is closer to the small *m* than is the case for pupils in general education. The pedagogical systems in general and vocational education are strongly differentiated. Mathematics in vocational education is completely embedded in a modular system, while mathematics in general education is a separate course.

Due to this different pedagogical structure, pupils in vocational education are taught the core skills of mathematics in response to the demands of the environment and are connected to real live problems. These pupils are taught the small *m,* while pupils in general education are taught the introduction of the large *M* in a separate course with a curriculum that is strongly directed towards the performance of techniques. This brings us to three critical remarks concerning the curriculum of mathematics.

4. CRITICAL REMARKS

In this last section, we would like to embed the findings of our case study in a broader critical theory on mathematics education. We, therefore, have to make the connection between our findings and contemporary theory on 1) didactics (Bishop [1988a] 1997; Cohen and Lotan 1997), 2) ethnomathematics (Powell and Frankenstein 1997; Powell 2002; Setati 2002), and 3) values in mathematics education (Bishop 2002; Bishop 2006; Forgasz 1999; Morge 2005). In doing so, we embark on three critical remarks concerning the three characteristics of the screened curriculum. First we would like to criticize the didactics frequently used in classrooms. Second, we would like to widen the narrow use of the word *ethnomathematics.* Finally, we will reveal so-called value-free mathematics.

4.1 The Absolutist View on Mathematics has an Impact on the Didactics Used

Looking at the results of the curriculum screening, we did not say much about the way in which mathematics is taught. One could clearly anticipate the link between the content of the curriculum and the manner of teaching, however. In addition, it is not surprising that the didactics used in vocational education are completely different from that used in general education. The pedagogical systems used in general and vocational education are strongly differentiated. Mathematics in vocational education is completely embedded in a modular system, and project-based teaching commonly uses didactics. In general education, mathematics is a separate course and the didactics frequently used are rather technique-oriented, as we shall explain below. Due to this different pedagogical system, pupils in vocational education are taught the core skills of mathematics in response to the demands of the environment, and are connected to real world problems. A strong base for higher education is provided in general education but vocational education is directed to the practice of a profession.

With Bishop ([1988a], 1997, 7-10) we see three major areas of concern about the present state of mathematics teaching in general education, where the introduction of *M* mathematics is required by a society based on knowledge. He distinguishes between the small *m* and the large *M* of mathematics, where the large *M* stands for mathematics as a *Western* scientific discipline, and the small *m* stands for *universal* mathematical basic competencies such as counting, locating, measuring, designing, playing, and explaining.

In these terms, we can say that pupils in vocational education are taught a form of mathematics that is closer to the small *m* than is the case for pupils in general education. In the case of Flanders, the following criticisms of the didactics used are applicable to general education, which is 1) strongly directed towards the performance of techniques, 2) a more impersonal learning process, and 3) more dominated by textbooks.

4.1.1 The Performance of Techniques

The technique-oriented curriculum lays down mathematics as the performance of techniques without any reflection or interaction. It is based on the assumption that a top-down approach is superior. It therefore does not portray mathematics as a reflective subject, and there is less room for interaction during the educational process. For example, in Flanders secondary education there is no attention paid to critical reflection, nor to the historical, human, and cultural aspects of mathematics. Mathematics is

presented as if there exists one and only one mathematical system that is made up of absolute truths and certain knowledge. The deductive method provides the warrant for the assertion of mathematical knowledge. Hence, the curriculum is strongly directed towards the performance of techniques. Arithmetic is entrenched as the basis of the mathematics curriculum from the very beginning, with the first instruction occurring in kindergarden and during primary school. In secondary education, arithmetic is gradually developed to handle more and more complicated number systems (N, Z, Q, R, and C). From the first year of secondary education, pupils are taught algebra and geometry. Later on, for those who have succeeded, the gateway stands open to further delights such as statistics, functions, integrals, and differentials—depending on the level and type of education. The curriculum is fully loaded with those technical performances. Pupils drop out at different levels when the subject becomes too difficult or too meaningless. The Flanders secondary school system is characterised by the so-called *waterfall system* where pupils drop to a *lower* level when they fail—*lower* in the sense that it has less status at school and in society in general. Mathematics plays a prominent role—more than sciences—as a gatekeeper in society. It is used as a selection device for entry to higher education or employment. The mathematical skills that are tested are not automatically related to the ultimate purpose (Bishop 2006). Mathematical skills, as such, are the selection device.

For those who succeed, and for the *smart* ones (e.g., the curriculum for the 'eight hours mathematics'), however, the curriculum includes room for moments to explore the broader context of mathematics. It is strange to see how teachers use the virtual mathematical Yahoo group to inform each other about what to do with those *free* hours (confirming their uneasiness with the topic, as they are successful mathematicians).

The question remains, "Why do we transform pupils into technique handlers *par excellence,* without any understanding or critical awareness of why, how, and when to use these mathematical techniques? Why don't we enable the learner to develop a critical stance either inside or outside mathematics?" A technical-oriented curriculum reduces us to merely instructing and training pupils without any critical reflection.

4.1.2 Impersonal and Individual Learning Process

Within an impersonal learning process, the task for the learner is conceived as being independent of the personal interest of the learner. The main system of education, and especially mathematics education, perpetuates this idea, which is closely connected to the top-down approach. There is no time (needed) for interaction in classroom or for experience-based learning. The

teacher is seen as the expert, and the learners are seen as empty barrels. It may well be that the teacher does recognise the personal interests of the learners, or that pupils may have philosophical interests in mathematics, but it has nothing to do with mathematics education per se. Reflections on mathematics are not a part of mathematics education—at best it can be a topic in a philosophy course.

In line with Van Moer (see Van Moer, this volume), we can say that mathematics is often presented as a purely deductive science, using a technique-oriented curriculum. The rules must be learned, the procedures accepted, and the skills practised. It doesn't matter what the learner brings to the situation. The mathematical result is, and shall always be, the same. Learners are not regarded as individuals but as generalized learners. Even if universal mathematical truths exist (e.g., the universality of *m* mathematics), that does not mean that mathematical education should ignore the learner's individuality. As Batanero and Díaz (see Batanero and Díaz, this volume) argue, the emphasis on the individual process of learning and understanding probability can give us an insight into the problems and the struggle of learners. Moreover, some *false* or *mistaken* answers can show a truly logical reasoning behind them. Especially in multicultural and multilingual classrooms, a personalized learning process should enable more pupils to succeed. It implies that presenting mathematics as an impersonal object to be transmitted in a one-way communication is a matter of priorities. To do differently—to leave room for interaction, for *views*, and for discussion—is the same.

Much research about interaction in the classroom has been done by Cohen and Lotan (1997). They developed a didactic called Cooperative Learning in Multicultural Classes (CLIM). CLIM is a cooperative didactic based on Complex Instructions (CI), as invented by Elisabeth Cohen at Stanford University, to foster the participation of inner city children in the educational process (Cohen and Lotan, 1997). The method basically consists of giving a challenging task to a group of pupils that has to be solved by them by following a highly structured procedure. Each of the pupils was assigned a specific role. In total, there were five roles and each pupil (at a given point in time) takes on each role within the execution of the task. In this way, pupils become dependent upon each other. They will have to draw as much as possible on each other's capacities to fulfill the task. Because of this principle of rotating roles, every pupil is confronted with varying duties, among which there is always at least one that he or she is able to perform well. In this way, they learn by doing and simultaneously learn the strong and weak points of themselves and the others. The tasks are carried out by the pupils only. The teacher is the initial organizer and functions mainly as a mediator. This means he or she will intervene, only when asked by the pupils

to do so. In this kind of teaching, the pupil is active and a producer of knowledge. This implies a changed role for the teacher, who is not immediately evident. When pupils participate in the process of building knowledge, the way the session develops is uncertain and partly unknown (Morge 2005). This creates a rather uncomfortable situation for the teacher, who is supposed to capture absolute mathematical knowledge by using the structured performance of the textbooks. This brings us to the final aspect of the didactics used in general education, according to the textbooks.

4.1.3 Dominance of Textbooks

In Flanders, there are more than a hundred textbooks with a companion workbook available only for mathematics in secondary education. Each level at each school system has its own books. Textbooks are firmly based on the curriculum as developed by the different school systems, and based on the curriculum as developed and recognized by the government. A teacher does not need to elaborate on the curriculum. Using the textbook, the teacher can be sure that he or she is teaching in the *right* way and that he or she will reach the attainment targets with his or her pupils. Teachers are controlled by textbooks and therefore are prevented from knowing their learners and from helping them. The dominance of the textbook embodies the top-down approach. The mathematical material for exams is written down in textbooks and provides certainty both for teachers and learners. The companion workbooks offer the possibility to practise the technical procedures at length. The learners have to sit quietly, listen carefully, and receive instruction without hesitation, and they have to practise and give the right answers (which are sometimes given in a special companion book for teachers). Here, the question remains how to motivate pupils in such a sterile educational process.

Teachers should be educated not to be dependent on the textbook. Teacher training needs to enable the teacher to control the material and not vice versa. The responsibility for teaching lies with the teacher and not with the text. If teachers need to personalize the educational process, they need to work critically with the available material and include the interests and experiences of the learners. They can personalize the teaching process by taking into account the diversity in the classroom. This is a central aspect to be considered and dealt with if we wish to introduce ethnomathematical praxis in Western curricula.

4.2 The Role of Ethnomathematics

In this section we will explaine the role of ethnomathematics in Western school curricula to overcome social stratification.

4.2.1 The Notion of Ethno

In the early 1980s, the word *ethnomathematics* was used for the mathematical practices of *non-literate* peoples. Marcia Ascher (a mathematician), and Robert Ascher (an anthropologist), (Ascher and Ascher [1986] 1997), used the term *non-literate* to counter the outmoded usage of the term *primitive*. Their work has provided detailed evidence of sophisticated mathematical ideas among non-literate peoples—ideas akin to, and as complex as, those of modern *Western* mathematics.

It was D'Ambrosio, a Brazilian mathematician and philosopher of mathematics education, and *the intellectual father of the ethnomathematics program* who presupposed "a broader concept of *ethno*, to include all culturally identifiable groups with their jargons, codes, symbols, myths, and even specific ways of reasoning and inferring." (D'Ambrosio [1985] 1997, 17; Powell 2002, 18; Vithal and Skovsmose 1997, 138). The word *ethno* should be understood as referring to cultural groups, and not as the anachronistic concept of race. Even the concept of *culturally differentiated groups* is no longer reserved for non-literal people. The use of the word is broader now and refers to heterogeneous groups, even within a Western classroom (Cohen and Lotan 1997). The notion of *ethno* refers to the ethnic, national, racial group, the gender, the professional group, and to the cultural group defined by a philosophical and ideological perspective (Powell 2002, 19). In short, we can say that the notion of *ethno* refers to diversity within mathematics and within mathematical teaching. Within ethnomathematics, scientists are currently collecting examples and data on the practices of culturally differentiated groups that are identifiable as mathematical practices. Within this field of research, we also have to look further than classical research on the mathematical practices of non-literate peoples. Even in a highly literate group, there can be a great difference between the mathematical practices of the different cultural subgroups. This also seems to be an interesting subject within research on ethnomathematics (Vithal and Skovsmose 1997, 143-135).

Obviously, we are living in a multicultural society. Even Western classrooms are characterised by diversity, by diverse mathematical practices, and by a diverse use and understanding of mathematical concepts. Each pupil is a cognizant being that functions within the language and interpretative codes of his or her sociocultural group. The challenge for

teachers nowadays is to deal with this diversity in such a way that it gives pupils equal chances. This brings us to the question, “What should the role of ethnomathematics be in Western classrooms?”

4.2.2 Dealing with Diversity

A central question within ethnomathematics today is whether the philosophical, ideological, and discursive norms of ethnomathematics differ from those of mathematics (Setati 2002). Many definitions suggest that ethnomathematics is a special type of mathematics. As Vithal and Skovsmose (1997, 142) have argued, most of the definitions or descriptions of ethnomathematics feature the term mathematics in various ways. Mathematics has to do with mathematical knowledge, with mathematical ideas, with mathematical activities, and with mathematical practices. There is no fundamental difference between mathematics and ethnomathematics. We could say that ethnomathematics gave the floor to critical voices, to other minority voices, and to *different voices* in mathematics education (Gilligan 1982). The central question within the discussion on the actual status and meaning of ethnomathematics is not whether ethnomathematics is its own discipline. Ethnomathematics is the way in which mathematics is handed down to pupils and this involves two aspects of diversity. First, there is the aspect of diversity within the content of the curriculum. Second, there is the aspect of diversity in the classroom, where the central question is “What do the children bring to school?” (Graham 1988) Referring to the definition of numeracy given by the OECD/PISA, we can say that we should teach mathematical literacy to all pupils.

> “Mathematical literacy is an individual’s capacity to identify and understand the role that mathematics plays in the world, to make well-founded judgements and to use and engage with mathematics in ways that meet the needs of that individual’s life as a constructive, concerned and reflective citizen.” (OECD 2003)

This definition is clearly not a neutral definition. Instead, it includes values as a fundamental part (Bishop 2006). Even if we say that we are now trying to teach mathematics to all pupils, we use a political statement—a statement which is fundamental to the political philosophy of the emancipation of all people within a democratic school system. Trying to teach mathematics to all pupils implies that we have to deal with diversity within the content of mathematics and with the diversity found in the classroom (Vithal and Skovsmose 1997, 145). This way of teaching mathematics is ethnomathematics.

The classical ethnomathematical enrichment of a curriculum is seen as "investigating the ethnomathematics of a culture to construct curricula with people from that culture, and by exploring the ethnomathematics of other cultures to create curricula so that people's knowledge of mathematics will be enriched." (Powell and Frankenstein 1997, 249). The questions then remain: "How can we enrich a Western curriculum? How can we apply this interesting view to Western school curricula? Moreover, why should we enculture mathematics in education while we (in Flanders) have nearly the best school results for mathematics on the international level, as presented in the results of the TIMSS[6] report." Flanders has the best results in Europe and is in fifth place on the international level (after Singapore, Hong Kong, Japan, and Chinese Taipei) for fourth-grade students. For eighth-grade students, it has the best results in Europe and is in sixth place on the international level (after Singapore, Republic of Korea, Hong Kong, Chinese Taipei, and Japan). These results are nearly the same as the results presented by the PISA[7] report. Flanders is in third place in Europe and reaches eighth on the international level (after Hong Kong, Finland, Korea, Netherlands, Liechtenstein, Japan, and Canada) for 15-year olds in school. Even if we have such a high level in international ranking, we have to reflect on it critically.

At the same time, we have to recognize the social stratification in education and the fact that our school system perpetuates social inequality in society. Social inequality becomes visible in primary education and increases in secondary and higher education (Nicaise 2003). Mathematics selects pupils at hierarchical levels. The more hours that are reserved for

6 The Trends in International Mathematics and Science Study (TIMSS, formerly known as the Third International Mathematics and Science Study) provides reliable and timely data on the mathematics and science achievements of U.S. fourth- and eighth-grade students, compared to that of students in 45 other countries. Offered in 1995, 1999, 2003, and 2007, TIMSS provides trend data on students' mathematics and science achievements from an international perspective. TIMSS is conducted under the auspices of the International Association for the Evaluation of Educational Achievement (IEA) at http://nces.ed.gov/timss.

7 The Programme for International Student Assessment (PISA) is an internationally standardized assessment that was jointly developed by participating countries and administered to 15-year-olds in schools. PISA is a three-year survey in the principal industrialized countries. The survey was implemented in 43 countries in the first assessment in 2000, in 41 countries in the second assessment in 2003, and in 57 countries in the third assessment in 2006. It assesses how students near the end of compulsory education have acquired some of the knowledge and skills that are essential for full participation in society. PISA is sponsored by the Organization for Economic Cooperation and Development (OECD). While PISA is on a 3-year cycle, TIMSS is on a 4-year cycle (http://www.pisa.oecd.org).

mathematics in the curriculum equal a higher status in the education structure and the more chances there are to improve and succeed in higher education. This kind of educational system is characterized by contradiction and perpetual motion.

On the one hand, it is the educational system that makes students feel that mathematics is important to study, and that it is important to succeed in higher education and to achieve a higher status in society. On the other hand, it is the very same educational system that makes students fail and that perpetuates social stratification. In Flanders' school system, the so-called *eight hours of mathematics* is reserved for the elite pupils whose teachers dream of teaching. It seems to be a contradiction that the system that creates the need is the same system that fails to satisfy that need.

Besides the contradiction within the system of mathematics education, there exists a perpetual motion where the *successful* never question their mathematical knowledge or their mathematics education. After all, there is no need to do so if you are successful. Furthermore, you need to be a rather successful student to become a teacher in mathematics. Hence, teachers do not really deal with the frustration of the *losers* in mathematics.

If we try to generate a democratic school system, it is a challenge in teacher training to try to teach mathematics to everyone. If mathematics education is about helping people to relate better to their environment, then it is clearly failing in this task.

We would like to emphasize the importance of using culture within the learning process to include—broadly speaking—societal as well as personal experiences of pupils to make mathematics learning more effective for *all* pupils. This is the way in which ethnomathematics can enrich Western school curricula. It is the way in which teachers can develop the self-confidence of pupils and replace their previous feelings of alienation towards traditional *school* mathematics. By doing this, mathematics can become a matter of intellectual stimulation instead of a dead-boring, pointless game. If we are now in a period of trying to teach mathematics to all, then we need to be critical of traditional practice and the current curriculum.

So far, we have criticized didactics. Now we move on to criticize the so called value-free curriculum.

4.3 Revealing Values in Mathematics Education

The mathematics that is taught is presented as being value-free. Mathematics is dehumanised, depersonalised, and decontextualised. It seems to be stripped of all historical and cultural phenomena to retain its *purity* and to avoid its disastrous contingencies. In line with this conception of mathematics, teachers are unaware of the values that they are teaching

within mathematics education. There even seems to be a lack of reference to any teaching of values by either curricula documents or textbooks. Teachers have no vocabulary to articulate implicit and explicit values. Some research is done to overcome this lack (Bishop 2002).

In his article, Heefer (see Heefer, this volume) gives some arguments against the view that mathematics offers absolute truth and explaines that it is not so much a lack of knowledge, but rather the static and unalterable mode of presentation of concepts in the mathematics curriculum that contributes to this misconception. He argues for the integration of the history of mathematics within the mathematics curriculum as a way to teach students about the evolution and context-dependency of human knowledge. Moreover, to enculturate our learners properly, we should be conscious of inherent values as a first step, while in the following steps we should make those values explicit. The implicit philosophy should become explicit so that we are conscious of what we teach.

In literature, there are several sets of values questioned. Alan J. Bishop ([1988a] 1997, 60-82) presents different sets of ideals and values associated with mathematics. He differentiates between six categories in three dimensions: the value of rationalism and objectism (the dimension of ideology), the value of control and progress (the dimension of sentiment), and the value of openness and mystery (the dimension of sociology). He makes the difference between four components of culture: the ideological, the sociological, the sentimental, and the technical component, where the technological factor is the basic one and all others are dependent upon it. Later on, he elaborates on his research on values in mathematics and science education (Bishop 2006). He had to, therefore, make a small change in the initial framework used for the interpretation of values in mathematics education. The value of objectism had to be recast as empiricism to accommodate scientists. The highly empirical nature of science means that it has many more value aspects than does mathematics, due to the experimental and observational activities of science. The model suggests that there are strong similarities in the values of science and mathematics. So, it comes as no surprise in the research findings that empiricism and rationalism are the most important values for both mathematics and science. However, interesting and, in terms of education, revealing different values are also represented. For secondary education especially, we can see an important difference between the rankings for mathematics and science and between empiricism and rationalism, where rationalism ranks much higher in mathematics than it does in science.

Revealing values in science education seems to be important. Findings of quantitative research that has been done in the field of science education (physics) shows the relationship between the teacher's beliefs and the way

he or she manages the conclusion of the pupils' answers (Morge 2005). More research on this topic is required to extrapolate this relationship in the case of mathematics.

In addition, Paul Ernest inquires about the inherent values in mathematics. Abstract is valued above concrete, formal above informal, objective above subjective, justification above discovery, rationality above intuition, reason above emotion, general above particular, theory above practice, the work of the brain above the work of the hand, and so on (Ernest 1991, 259).

Much research has been done on the issue of gender and mathematics. Since 1976, the Fennema-Sherman Mathematics Attitude Scales (MAS) have been frequently used to measure attitudes towards mathematics. Boys always stereotyped mathematics more strongly as a male domain than girls did. Mathematics as a male domain still remains a relevant variable in explanations for persistent patterns of gender differences in the learning of mathematics (Forgasz 1999).

In a highly original paper, Suzanne Prediger (2006) emphasises the value of coherence and consensus. She answers the question of why mathematical theories are always made or remade coherent. The prevalent view of mathematics is that when the mathematic community is convinced that inconsistencies may not appear, participants will make great efforts to remove them whenever they do appear.

Unfortunately, we cannot elaborate here on all these values. We would like to mention two which are, in our opinion, the most interesting: rationalism and objectivism.

Before we present our case, we would like to clarify that we do not argue against rationalism and objectivity. We would not like to be suspected of radical relativity, of subjectivity, and of course not of promoting the occult sciences or forms of divination. But over the years, we have learned that it can be fruitful to be explicit from time to time. All values have their power on the one hand and their limitations on the other.

4.3.1 The Value of Rationalism

Rationalism is undeniably at the heart of mathematics. It guarantees its power and authority. This value is ranked as the highest value in mathematics by teachers (Bishop 2006). Rationalism has to do with certain knowledge, with deductive reasoning, with logical truth, completeness and consistency, as opposed to tradition, religion, personal beliefs and biased knowledge. Rationalism has become more and more important in our Western culture and has largely influenced the organization of our complex society, our view of theworld, and our view of human beings. Rationalism is

present everywhere in our culture, including our own personal life. Arguments are accepted and tolerated, while a lack of logical coherence in an argument is not. What needs to be explained in education is the fact that it is not the tangible world of material objects that is logical, nor people or things that are rational. It is mathematical explanations and representations that are rational, logical, and, abstract. This is one perspective of how to explain and represent the world. We must, however, accept the power of abstracting and operating with ideas. In mathematics, it is a particular way of theorising about phenomena that is valued.

For pupils to appreciate rationalism, it is necessary to make them aware of what it is to explain, to abstract, and to theorize—all tools that we need to solve problems. Without understanding this, the language and symbols of mathematics will be totally meaningless to our pupils.

The argument for rationalism does not imply a negative connotation of emotions. It has nothing to do with the classical opposition between rationality and emotions. In some areas, we need the application of mathematics, while in other domains we need our emotions. Pupils have to understand this and know how to deal with it.

4.3.2 The Value of Objectivism

At the dawn of modern science, it was Galileo who wrote "the book of nature is written in the language of mathematics" and Descartes also saw mathematics both as the language in which nature is written and as the method to deduce our knowledge about nature.

This philosophical idea about science became the core of our modern conception of science and was further generalized from then on. Moreover, the idea of the mathematization of the world, of grasping it with absolute certainty and hence with the highest degree of objectivity, became a challenge not only for the so-called *hard* sciences but also for the human or *soft* sciences.

The core story of present-day sciences seems to be a story of neutrality, without any political connection. Nevertheless, it is a fact that taking an option for a method that guarantees the highest certainty and objectivity implies constraints on their objects of knowledge.

So, if the method determines the objects that are knowable, and if the method actually *produces* these objects of knowledge, then we can claim that it indeed produces objective knowledge, though definitely not neutral knowledge. Any use of a particular method implies choices and, in general, choices having social relevance. Therefore, we can indeed call such socially relevant choices *political acts* (in the broad meaning of the term).

Little research has been conducted regarding applied mathematics and its political, social, and ethical impact. The most obvious relationship seems to be the connection between mathematics and war, where mathematicians have lent their services and expert knowledge in the furtherance of war. Another much less evident example, as we have seen, is the way in which mathematics is handed down in education, and how it is taught—including a set of values, both implicit and explicit.

What, then, is the political aspect of the story of neutrality and objectivity? Basically, it is the proliferation of one perspective that elevates itself above all others—namely the *objective* perspective—and moreover that this claims neutrality. The choice for an objective representation of nature is presented as neutral in the sense that within this perspective one removes the impact of subjectivity (to the best of one's capabilities), as well as all needs and interests (those of objectivity itself excepted). It is *not* neutral, however, to make such a choice concerning the way things are represented—a choice concerning how to *epistemize* the world—in this case, in an abstract way by isolating things and stripping them of various *variables* aimed at grasping nature in universal and immutable laws and representing it in a formal framework. At this point mathematics enters the picture, i.e., as *the* language to objectify, to abstract, and to isolate things. This is also where mathematics becomes both neutral and political—political in the sense that one makes a particular *choice*, however implicit, to represent nature in an objective way, sedimented in universal laws, and a choice that is made without any ideological, social, or political argumentation. While Galileo may very well have preached that *the* book of nature is written in the language of mathematics, we contend that, contrary to this, he has referred to only one of the possible books with surely many more to be written (François and De Sutter, 2004).

Let us repeat that we believe that teachers of mathematics should be conscious of what they teach. Therefore, any implicit philosophy, including hidden values, has to be made explicit. By doing this, we can enculturate and humanize the curricula of mathematics—once again creating room for ethnomathematical praxis in Western curricula.

5. CONCLUSION

A cultural, historical, and philosophical perspective of mathematical knowledge implies much more than merely teaching children to do mathematics. Education, as one of the most important societal institutions that hands down knowledge and reflections on mathematics, has to go

beyond the stereotype of mathematics as *coldly rational* without any connection to the real world.

If we want to give mathematics its proper social status, we have to teach how mathematics is embedded in our culture and our history—how other kinds of mathematics are embedded in their cultures and their histories. Mathematics presented as a cultural practice implies a view on mathematics with room for diversity. At the same time, mathematical practices can differ from one pupil to another, even if they are sitting next to each other in the same classroom. If we want to teach our pupils to become critical citizens, mathematics has to be taught as a way of *knowing*, rather than as a series of *automated* techniques.

Moreover, if it is our aim (as it us for us now), to teach mathematics to all pupils, we need to criticize traditional practice and the current curriculum. Even in those countries with the highest levels of mathematical knowledge and mathematical literacy, school systems perpetuate social inequality that increases during a school career. This is not registered, however, as it seems that mathematical education is value free. Opposing this view, we argue that we do transmit values, maybe unconsciously, but certainly implicitly and uncritically. We need to make these inherent values explicit so that teachers are conscious of what they teach.

In the first place, we argue for the implementation of philosophy in the curriculum of mathematics now that we are in an era of trying to teach mathematics to all pupils, and in the second place we argue that as we are now living in a multicultural and global world we need to enculturate our learners properly. We are, therefore, challenged to make mathematical education a more humanized and encultered process.

REFERENCES

Ascher, Marcia and Ascher, Robert [1986] (1997). Ethnomathematics. In Powell, Arthur B. and Frankenstein, Marilyn (Eds.) (1997). *Ethnomathematics, Challenging Eurocentrism in Mathematics Education*, State University of new York Press -SUNY, Albany.

Bishop, Alan. J. [1988a] (1997). Mathematical Enculturation, A Cultural Perspective on Mathematics Education. *Mathematics Education Library,* Vol. 6, Kluwer Academic Publishers, Dordrecht / Boston / London.

Bishop, Alan J. (1988b). The Interactions of Mathematics Education with Culture. *Cultural Dynamics, An International Journal for the Study of Processes and Temporality of Culture*, Volume I. 2, 145 - 157.

Bishop, Alan .J. (2002). Research Policy and Practice: The Case of Values. In Valero, Paola and Skovsmose, Ole (Eds.) (2002). *Mathematics Education and Society. Proceedings of the Third International Mathematics Education and Society Conference* (MES 3), second edition, 2 Vols. Centre for Research in Learning Mathematics, Helsingor, Denmark, 227-233.

Bishop, Alan J.; Clarke, Barbara; Corrigan, Debbie and Gunstone, Dick (2006). Values in Mathematics and Science Education: Researchers' and Teachers' Views on the Similarities and Differences. *For the Learning of Mathematics, An International Journal of Mathematics Education.* Vol. 26, num. 1,7-11.

Cohen, Elizabeth G. and Lotan, Rachel A. (Eds.) (1997). *Working for equity in heterogeneous classrooms: sociological theory in action.* New York: Teachers College Press.

D'Ambrosio, Ubiratan [1985] (1997). Ethnomathematics and its Place in the History and Pedagogy of Mathematics. In Powell, Arthur B. and Frankenstein, Marilyn (Eds.) (1997). *Ethnomathematics, Challenging Eurocentrism in Mathematics Education*, State University of new York Press -SUNY, Albany.

Descartes, René [1684] (1984-1991). Regulae ad directionem ingenii (Rules for the Direction of the Mind). In Cottingham, R.; Stoothoff, D. Murdoch, and A. Kenny (Eds. and Trans.) *The Philosophical Writings of Descartes*, Vol. 1, Cambridge University Press, Cambridge.

Ernest, Paul [1991] (2003). *The Philosophy of Mathematics Education, Studies in Mathematics Education*, RoutledgeFalmer, London.

Ernest, Paul, (Ed.) (1994). *Constructing Mathematical Knowledge: Epistemology and Mathematical Education, Studies in Mathematics Education Series* 4, London: The Falmer Press.

Ernest, Paul (1998). *Social Constructivism as a philosophy of Mathematics*, (SUNY series in science, technology, and society, editors, Sal Restivo and Jennifer Croissant and SUNY series, reform in mathematics education, editor, Judith Sowder) Albany: State University of New York.

Forgasz, Helen J.; Leder, Gilah C. and Gardner, Paul L. (1999). The Fennema-Sherman Mathematics as a Male Domain Scale Reexamined. *Journal for Research in Mathematics Education.* Vol. 30, No. 3, 342-348.

François, Karen and De Sutter, Laurent (2004). When Mathematics Becomes Political. Representing (Non)Humans. *Philosophica*, Vol. 74, 123-138.

Gilligan, Carol (1982). *In a Different Voice; Psychological Theory and Women's Development.* Harvard University Press.

Graham, Beth (1988). Mathematical Education and Aboriginal Children. In Bishop, Alan, J. (Ed.) (1988). *Mathematics Education and Culture. Educational Studies in Mathematics.* 19, Kluwer Academic Publihers. The Netherlands, 119-135.

Licap (1997). *Leerplan Secundair Onderwijs, Wiskunde, Eerste Graad*, Licap – Brussel.

Morge, Ludovic (2005). Teacher-Pupil interaction: A Study of Hidden Beliefs in Conclusion Phases. *International Journal of Science Education.* Vol. 27, No. 8, 935-956.

Nicaise, Ides (2003). *Cijferboek sociale ongelijkheid in het Vlaams onderwijs, Een verkennend onderzoek op de Panelstudie van Belgische Huishoudens*, Onderzoek in opdracht van de Vlaamse minister van Onderwijs en Vorming, in het kader van het programma (Steunpunten voor Beleidsrelevant Onderzoek, LOA-rapport nr. 10, Leuven.

OECD (2003). *Assessment Framework – mathematics, reading, science and problem solving: knowledge and skills.* Paris: OECD.

OECD (2004). *First Results from PISA 2003 - Executive Summary.* Paris: OECD.

OECD (2004). *Learning for Tomorrow's World - First Results from PISA 2003.* Paris: OECD.

Powell, Arthur B. and Frankenstein, Marilyn (1997). Ethnomathematical Praxis in the Curriculum, in: Powell, Arthur B. and Frankenstein, Marilyn (Eds.) (1997). *Ethnomathematics, Challenging Eurocentrism in Mathematics Education*, State University of new York Press -SUNY, Albany.

Powell, Arthur B. (2002). Ethnomathematics and the Challenges of Racism in Mathematics Education. In Valero, Paola and Skovsmose, Ole (Eds.) (2002). *Mathematics Education and Society. Proceedings of the Third International Mathematics Education and Society*

Conference (MES 3), second edition, 2 Vols. Centre for Research in Learning Mathematics, Helsingor, Denmark, 17-30.

Prediger, Susanne (2006). Mathematics — Cultural Product or Epistemic Exception? In B. Löwe, V. Peckhaus, T. Räsch (Eds.) *Foundations of the Formal Sciences IV, The History of the Concept of the Formal Sciences,* King's College, London, 217-232.

Setati, Mamokgethi (2002). Is Ethnomathematics = Mathematics = Antiracism? In Valero, Paola and Skovsmose, Ole (Eds.) (2002). *Mathematics Education and Society. Proceedings of the Third International Mathematics Education and Society Conference* (MES 3), second edition, 2 Vols. Centre for Research in Learning Mathematics, Helsingor, Denmark, 31-33.

Vithal, Renuka and Skovsmose, Ole (1997). The End of Innocence: A Critique of Ethnomathematics. *Educational Studies in Mathematics*. Vol. 34 Issue 2, 131-157.

THE CURRICULA

The extended list of the references of the curricula (about six pages), developed by the community and the own-developed curricula by the differing school systems are available but not included here. Interested readers are welcome to contact the author. E-mail address: Karen.Francois@vub.ac.be.

INTERLUDE 2

The preceding contribution is a rather general analysis of what we could call a paragon of a mathematics curriculum, if we were to use the international comparative reports as reference here. Attention was paid to the absence of an explicit philosophy of mathematics and to the technical-mathematically high-level content of the curriculum. At the same time, it was shown how at a micro level, too, the segregation between strong pupils and weak pupils comes into play. At a macro level, one can frankly state that the Flemish mathematics education shines. However, if we take a closer look, we notice that the selfsame segregation mechanisms that appear worldwide repeat themselves at the local level. How are we to fight this segregation? How can we succeed in including all children in mathematics education? This not only requires a didactic skill from the teacher, but also a vision of mathematics education as something that needs to be more than just the technical-mathematically oriented exercises. A surplus value for mathematics education can be the philosophical reflection, not just to involve all pupils in the learning process, but also to bring to the classroom the genuinely philosophical items relating to mathematics. As a teacher, it is crucial to acquire a trained eye for the many philosophical ideas that open-minded pupils have at times. This, too, calls for a corresponding didactic. The one-way traffic, where, the teacher knows everything and the pupils know nothing, precludes the possibility of getting to know the sometimes ingenious ideas of pupils. A teacher who advocates an emancipatory project will need to be open to interaction. In the next contribution, we are presented a number of examples of what classroom practice can be like. But first and foremost, the philosophical question is put forward about what a *philosophical reflection* actually is before moving on to an application of it. Stimulating philosophical reflections in mathematics education can be done in a very explicit, guided, and purposeful manner, but it can also *just happen* in class. This contribution presents examples of both forms, opening one's eye to the way in which, as a teacher, you can seize moments of learning during the mathematics lesson to reflect philosophically about a mathematical theme. As the examples show, philosophical reflections about mathematical themes can have a very playful preamble—like the story about *The Little Prince*—to end up with a rather heavy-looking philosophical debate about something like the *ontological status* of mathematical objects.

PHILOSOPHICAL REFLECTIONS IN MATHEMATICS CLASSROOMS

Chances and Reasons

Susanne Prediger
Dortmund University, Germany

Abstract: "Mathematics in education: Is there room for a philosophy of mathematics in school practice?" That was the central question at the conference from which this volume grew. My answer to the question is yes—absolutely! In the article, I argue why and how philosophical reflections should be included in mathematics classrooms. The general ideas will be explained by three examples from classrooms.

Key words: Mathematical literacy, philosophical reflection, primary school

1. INTRODUCTION

"Mathematics in education: Is there room for a philosophy of mathematics in school practice?" That was the central question at the conference in May 2004 from which this volume grew.

It was nearly thirty years ago that several well-known German researchers in philosophy of mathematics and mathematics education posed the question of whether philosophy should be integrated into mathematics classrooms and how (cf. Otte 1977). The researchers extensively discussed the relationship between mathematics and philosophy and made suggestions for how to include philosophical considerations in mathematics education. But although all authors agreed on the importance of philosophical considerations in mathematics classrooms, they had to admit that classroom practices did not reflect this importance.

Nearly thirty years after these discussions, the question must still (or again) be posed in the same sense. Since then, many aspects of the philosophy of mathematics and classroom practices have changed. Whereas the suggestions of 1977 basically concentrated on classical philosophical issues (like questions for the foundations of mathematics, the ontology of mathematical objects, and the status of mathematical truths), the contemporary philosophy of mathematics has shifted focus. Nowadays, large parts of the philosophy of mathematics are not only concerned with questions about the foundations, but with descriptions and analyses of mathematical practices (Kitcher 1984, Tymoczko 1985), the relationship between mathematics and human beings (Davis/Hersh 1980), and socio-philosophical reflections on the role of mathematics for society (Restivo/Fischer/Van Bendegem 1993, Keitel/Kotzmann/Skovsmose 1993, and many others). Mathematics is not seen as an absolute corpus of well-founded knowledge anymore, but as "a human activity, a social phenomenon, part of human culture, historically evolved, and intelligible only in a social context" (Hersh 1997, 11).

Not completely independent of these developments in philosophy, classrooms practices have also changed since 1977. For this contribution, the most important changes can be seen in the new orientation towards realistic mathematics education (de Lange 1996) and the development of innovative classroom practices (where argumentation and communication play a greater role, as well as learning by discovery—see, for example, NCTM-Standards 2000).

In spite of all these changes, I suspect that a survey about the explicit role of philosophy in mathematics classrooms would not provide a better picture than it did thirty years ago. Karen François's investigation of the Flanders (Belgian) curricula gives a hint for this assumption (cf. François/Van Bendegem 2004). So, what might be the reasons that philosophical considerations could hardly find an adequate place in mathematical classrooms, although it has been claimed many times?

The main argument against philosophical reflections that we hear from practitioners is always the argument of too little time, caused by overloaded syllabuses. Beyond this argument, we find a few other reasons. The most important reason might be that universities still educate too many teachers who are not acquainted with philosophical reflections themselves (which is an enormous deficit at least in German teacher education). Additionally, there are not many convincing materials for classrooms that stimulate philosophical reflections. The only exception is the well-equipped field of mathematics and reality, as far as mathematical modeling and the evaluation of their chances and limits are concerned.

Especially the field of mathematical modeling, however, has shown that the development of material for classrooms alone might be a necessary, but not sufficient, condition for modeling being integrated into classrooms. Decades of experience show that even more important than the materials are the teacher's conceptions and beliefs about mathematics (Thompson 1984). Hence, it is the image of mathematics that proves to be the biggest obstacle for modeling being included in classrooms (Kaiser/Maaß 2007).

Without knowing any empirical studies on obstacles of integrating philosophical reflections in the courses, I suspect that the situation might be similar. Philosophical reflections do not significantly appear in classrooms because they do not fit with many teachers' beliefs on mathematics learning because:

- philosophical reflections are often considered too difficult or only possible if the learning of the pure subject is completed
- philosophical reflections are considered *additional*, not integral part of mathematics
- philosophical reflections go against the traditional conception that mathematics education should be restricted to inner mathematical concepts, theorems and problem-solving procedures.

In view of all these obstacles, this article will not restrict itself to some suggestions of how to integrate philosophical reflections into mathematical classrooms. Instead, it aims to develop the thesis that for an adequate understanding of mathematics and an adequate specification of aims for mathematics education, philosophical reflections *must* play a prominent role in the learning process. And if they do, this is not only an additional (difficult?) learning content, but they can even enrich the mathematical learning in the narrowest sense. In order to substantiate this thesis, different aspects must be explained: What exactly is meant by philosophical reflections? (Section 2), What is the understanding of mathematics and mathematics education behind that? (Section 4) How can suitable reflections be stimulated? (Section 3).

2. WHAT DOES PHILOSOPHICAL REFLECTION MEAN?

To clarify the term *philosophical reflection*, we can start from Roland Fischer's definition of philosophy as reflecting discipline.

"[...] I understand philosophy as the reflecting discipline (Reflexionswissenschaft) with respect to being, perceiving and acting which has not set itself any boundaries" (Fischer 1982, 198, my translation).

This understanding of philosophy implies that including philosophy in mathematics classrooms is definitely not a matter of teaching about classical philosophers and existing philosophical theories. Instead, it should focus on *doing philosophy* in the sense of *reflecting philosophically*. This shift to the activity itself can well be expressed by creating the verb *philosophize*.

Many mathematics educators have—for various reasons—suggested that the learning process be enriched by deeper reflection (Pólya 1945, Steiner 1987, Cobb, et al. 1997, Neubrand 2000, Sjuts 2002, Prediger 2005, Lengnink/Siebel 2004, and many more). Reflecting alone, however, can not characterize a philosophical approach, because reflection can be concentrated on many different issues. In his article, *Reflecting as a Didaktik Construction*, Neubrand (2000) structured the complex and widespread field of suggestions for areas of reflection by specifying four different levels of reflecting on and speaking about mathematics.

> The level of the mathematician:
> "Speaking about mathematical subjects and problems themselves, for example, about the correctness of a proof, about the adequacy of the formulation of a definition, about logical dependencies, and so on."
>
> The level of the deliberately working mathematician:
> "Speaking about specific mathematical ways of working, their value and meaning, for example, about heuristic techniques in problem solving; about various modes of concept formation in mathematics, about specific mathematics methods like systematization, classification, or abstraction; about schemes and techniques of proof; and so forth [...]."
>
> The level of the philosopher of mathematics:
> "Speaking about mathematics as a whole with critical distance, for example about the roles of applications and their relation to mathematics concepts, about proofs as a characteristic issue in mathematics, and so on [...]."
>
> The level of the epistemologist:
> "Speaking about mathematics from an epistemological perspective, for example, about the characteristic distinctions between mathematics and other sciences, about the nature and the origin of mathematical knowledge, and so on [...]."
>
> (Neubrand 2000, 255f)

Obviously, most of the reflections in mathematics classrooms are (and should be) located on the *level of the mathematician* itself - for example, when open-ended problems or powerful tasks are explored and discussed in the class (Becker/Shimada 1997, Krainer 1993).

Figure 1: Levels of Reflection.

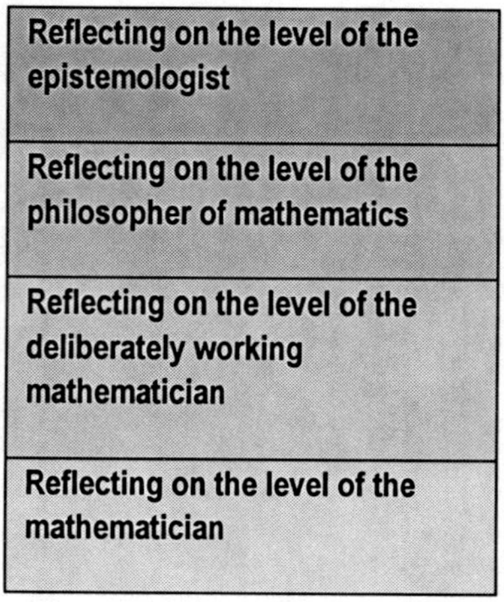

The most emphasized level of reflection in newer mathematics education literature is located on the *level of the deliberately working mathematician.* Especially all aspects of meta-cognitive activities (i.e., the thinking about your own thinking) are located on this level. They are emphasized by many researchers who are inspired by cognitive psychology (like Flavell 1979, Sjuts 2002). Classical problem solvers (like Pólya 1945, Schoenfeld 1992), however, have already emphasized their importance. Many empirical studies have shown that metacognitive awareness on the *level of the deliberatively working mathematician* can enhance learning processes in a significant way.

In contrast, the *level of the philosopher of mathematics* and the *level of the epistemologist*, being the natural locations of philosophical reflections, are often underemphasized in the theory of mathematics education, as well as in practice. Those are exactly the levels that are the focus of this article.

What are the current issues of philosophical reflections? Steiner has compiled the following list of subjects within philosophical thinking.

- "Questions of justifying mathematical knowledge (context of justification), especially the role of proofs in this context
- relations between pure and applied mathematics, methodology of mathematisation and modeling
- the role of problems and problem solving in mathematics
- the role of tools for representation and cognition
- the dynamics of genesis and development of mathematical concepts and theories and the role of proofs in this context
- relationship between justification, application and development."

(Steiner 1989, 47f, my translation)

In light of the developments in the philosophy of mathematics in the last twenty years since Steiner's article, this list must be appended, at least by the following aspects:

- issues of mathematical practices (for example, technical language and its purposes, the questions considered to be important, etc.)
- role of mathematics in society
- relation between mathematics and human beings

All these aspects can and should be issues for philosophical reflection at all stages of mathematics education. Although I am personally convinced that even primary students can start philosophizing about these issues, I will restrict myself to the secondary level (my professional field of work) here.

3. STRATEGIES FOR INITIATING PHILOSOPHICAL REFLECTIONS – THREE EXAMPLES

How is it possible to teach students to reflect? Neubrand has pointed out that reflecting cannot easily be taught. Nevertheless,

> "Reflecting is always a very personal task of the learner herself; that is, it is her own responsibility. But teachers can provide opportunities and the stimulation of reflection." (Neubrand 2000, 252)

Mathematics education research has made many efforts to develop materials and reflection-oriented tasks that stimulate reflections on the *level of the mathematician* and of the *deliberative working mathematician* (e.g., Sjuts 2002, Kaune 2006, Krainer 1993).

Even for reflections on a higher level, it is possible to formulate reflection-oriented tasks, as the following example from my Grade 5 (10-year old students) class shows.

Example 1: The Little Prince and the Numbers

My Grade 5 students worked on the task shown in Figure 2. Obviously, the role of mathematical means in everyday-communication can be reflected much deeper with older students, and convincing examples have been given by Lengnink (2002). The answers of my fifth graders, however, made me optimistic that reflection about the role of quantitative descriptions in our society can and should be initiated at every age.

Figure 2: The Little Prince—Task for Grade 5; one student's answer (10-year old).

The Little Prince, the Grown-ups, and the Numbers

The Little Prince wonders why the grown-ups love numbers. "When you tell them that you have made a new friend, they never ask you any questions about essential matters. They never say to you, 'What does his voice sound like? What games does he love best? Does he collect butterflies?' Instead, they demand: 'How old is he? How many brothers has he? How much does he weigh? How much money does his father make?' Only from these figures do they think they have learned anything about him. If you were to say to the grown-ups: 'I saw a beautiful house made of rosy brick, with geraniums in the windows and doves on the roof,' they would not be able to get any idea of that house at all. You would have to say to them: 'I saw a house that cost $20,000.' Then they would exclaim: 'Oh, what a pretty house that is!'" (De Saint-Exupéry 1943, Chapter 4).

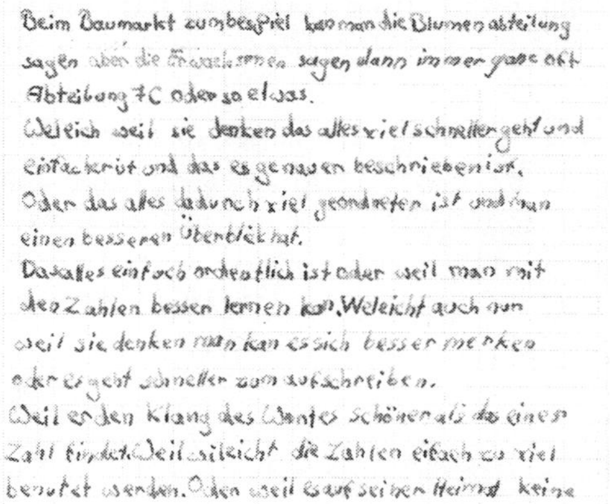
Beim Baumarkt zumbeispiel kan man die Blumen abteilung
sagen aber die Erwachsenen sagen dann immer ganz oft
Abteilung 7C oder so etwas.
Weleich weil sie denken das alles viel schneller geht und
einfacher ist und das es genauen beschrieben ist,
Oder das alles dadurch viel geordneter ist und man
einen besseren Überblick hat.
Das alles einfach ordentlich ist oder weil man mit
den Zahlen bessen lernen kan, Weleicht auch nur
weil sie denken man kan es sich besser merken
oder es geht schneller zum aufschreiben.
Weil er den Klang des Wortes schöner als den einer
Zahl findet. Weil vileicht die Zahlen eifach zu viel
benutzt werden. Oder weil er auf seinem Heimat keine
Zahlen gibt und es klingt so fremd oder eigenartig.
Ich stimme dem kleinen Prinzen zu, Weil ich finde
das wen ein Satz ohne Zahlen bessen klingt.
Und weil ich finde das es zu oft benutzt sind und das
es ohne Zahlen sehr viel genauer ist. Doch zum rechnen
und lernen finde ich die zahlen schon hilfsbereich
praktisch und schlau gedacht.
Ich finde das meinchen Leuten die zahlen niks bedeu-
ten und sie sich die menge gar nicht vorstellen können

Questions:

a.) Find another example in which grown-ups express things by numbers.

b.) Do you have an idea why the grown-ups like numbers so much? What benefits do we have by describing phenomena by numbers?

c.) Why is the Little Prince so critical of grown-ups' love for numbers? Do you agree with him? Why (not)? Can you also find an example where the description by numbers is problematic?

Translation of one student's answer:

a. For example, in the do-it-yourself store, we could say flower shop, but adults often say section 7c, or something like that.

b. Perhaps, because they believe that it is faster and easier, or that they

describe it more exactly. Or that everything is more ordered and you have a better overlook. Or that everything is just more structured, or that you can learn it better with numbers. Perhaps it is only because they believe they can remember it better or it is just faster to write down.

c. Because he prefers the sound of the words to the one of numbers. Perhaps because numbers are used too often. Or because there are no numbers in his homeland and they sound so strange and bizarre.

I agree with the Little Prince since I believe that a sentence without numbers sounds better. And I think that they are used too often and without numbers, it is more exact. But for calculating and learning, numbers are very helpful, practical and smart. I think that for some people, numbers do not mean anything and that they cannot imagine anything by it.

Although these examples show some possible ways to initiate reflections at the higher levels by prepared learning arrangements, we have to deal with the problem that many issues of philosophical reflections are by nature far away from students' interest and questions - at least at first sight. That is why it is not always easy to follow the important pedagogical imperative to connect all reflections to the students' prior experiences and their own questions.

One important idea in overcoming this difficulty is not to introduce all sequences of philosophical reflection by planning in advance. The more effective and pupil-oriented approach is to pick up situational chances for reflections that appear in the normal interactions in classrooms. Let me give an example of this *principle of situational reflection* (cf. Prediger 2004 for a detailed argumentation and further examples).

Example 2: Lisa, Matt and the Ontology of Mathematical Objects

This is a lesson about geometrically interpreted solutions of 2x2 linear equation systems in my colleague's Grade 9 classroom (15-year old students). Question: What happens if we have two equations of two parallel straight lines?

Lisa: Parallel lines never do meet; hence there cannot be a solution.

Matt: Oh yes, they do meet!

Lisa: No, they do not! Parallels cannot. Yours are perhaps not parallel. Look at these! [*draws parallel lines on the paper*]

Matt: They also meet, although very, very far away [*draws a meeting point on the table outside the paper*].

Having listened to this little controversy, the teacher mediated the controversy by adopting a meta-point of view and asked both students what ontological status they gave their lines:

Teacher: I have the impression that you are not talking about the same objects, aren't you? What exactly are those lines you are talking about? Where do they exist?

By means of this clarifying question, both students realized that for Lisa, parallel straight lines were theoretical constructs with idealized attributes (parallelism), whereas Matt focused on the drawn figures on the sheet of paper. Unlike the idealized constructs, drawn figures do have an existence in the real world, and with this ontological status, they indeed nearly always intersect.

By these considerations, the students were engaged in discussions about controversial ontological positions that are both well known in the philosophy of mathematics. Many questions can arise in such a discussion such as

- About what kind of objects can we state anything in mathematics?
- In which one are we interested?
- Which ontology is suitable for the original interest in linear equation systems?

Starting from the inner mathematical problem (do parallel lines have a meeting point?), this situation led into interesting philosophical considerations. The situational picking up of chances for reflections got its dynamics from the fact that the ontological question about the nature of lines here was not *another thing to learn* (as my students would say), but instead was an important tool for clarifying an ongoing mathematical controversy.

Like in this situation, philosophical reflections can often help to clarify different point of views. Another important approach to reflections is via questions for sense and self reflection (cf. Prediger 2005, where this idea is elaborated extensively). Again, an example shall illustrate the idea.

Example 3: Anne and the Missing Personal Access to Algebraic Calculus

While solving algebraic equations in Grade 10, my 16-year old student Anne asked me in a frustrated mood: "What have all these transformations got to do *with me*?" Anne was achieving quite well in the technical part but she was searching for *personal access* to the rules and techniques. My first

answer with the typical examples for applications (that can show the practical use of solving equations) could not convince her. So, we continued by discussing why she considered transforming equations not to be connected with her own person. In this way, we could approach the core of the problem: “When I solve equations, I feel like a machine. I do not even have to start real thinking.”

Following this idea of a machine, we reached an interesting place: We realized that it is an important characteristic of algebraic transformations that we can do them without *thinking*—i.e., without any interpretation of the syntactical steps. They can indeed be drawn mechanically since they are rule-guided and independent of any meaning in a concrete context. That is why Krämer talks about *symbolic machines* and describes the historic development as a “long-going and difficult history of the mechanical use of symbols, a history in which we learned to behave like machines when operating with symbols” (Krämer 1988, 4, my translation). In a certain way, this mechanical calculus is dehumanized and hence there is hardly any personal access possible. Exactly this characteristic, however, offers the very important opportunity to discharge our thinking.

Anne could experience this discharge directly when we reconsidered a geometrical question (concerning changes of areas) that we had solved by algebraic equations. We experienced that it is theoretically possible to interpret all the algebraic transformations in the geometrical context, but it is much more difficult to do so. Now, she began to value the possibility of discharge as a strategy of extending her own thinking. At the same time, Anne took enormous comfort in the knowledge that it took a long time in the history of algebra to develop this dehumanized calculus (Personal experience in my class).

Unforeseen in advance for me as the teacher, the frustration about solving algebraic equations and the search for making sense of it took a fruitful development in this situation. Via a self reflection (“What exactly is it that I do not like in transforming equations?”), Anne came to the realization of a fundamental idea of mathematics—algorithmizing —or, more generally, the rule-guided operating without interpretation. With reference to the concrete example of an algebraic equation, Anne has approached the philosophically important question about the relationship between human beings and mathematics and how it is threatened by its mechanization (see Prediger 2004 for this philosophical question).

Like in this example, the *search for making sense* of mathematical contents has often proved to be an excellent starting point for reflections that can (but not necessarily) become deeply philosophical. That is why Fischer has repeatedly underlined the role of negotiations about individual sense

constructions for mathematics learning (Fischer/Malle 1985, 9-20; Fischer 2001).

The most important precondition for applying the described principle of situational reflections is the teacher's awareness of the reflective potential underlying a situation. If this awareness is not well developed, teachers will miss the chance to exploit potentially reflection-rich situations as they were briefly sketched in the two examples. That is why this article does not want to stop with giving ideas about how to initiate philosophical reflections in the classrooms. It is even more important to make explicit the underlying positions about the aims of mathematics education and the understanding of mathematics.

4. UNDERLYING POSITIONS ABOUT MATHEMATICS (EDUCATION)

Philosophical reflections are not an arbitrary part of the learning process that can be exchanged by any other content. They belong to the core of *Bildung* as Hartmut von Hentig has emphasized.[1]

> "Literacy [Bildung] is a state of mind, the result of a contemplative way of approaching principles and phenomena of the own culture." (von Hentig 1980, 6, my translation)

This understanding of literacy is of great relevance also for mathematics. We can read this from the conception for mathematical literacy as it was defined by international consensus in the normative framework of PISA (Programme for International Student Assessment) implemented by the OECD (Organization for Economic Co-operation and Development, see OECD 1999).

> "Mathematics Literacy is an *individual's capacity to identify and understand the role that mathematics plays in the world,* to make well-founded mathematical judgments and to engage in mathematics, in ways that meet the needs of that individual's current and future life as a constructive, concerned and reflected citizen." (OECD 1999, italics added)

The PISA framework has been under critique because, in many test items, this deep and challenging understanding of mathematical literacy has

[1] *Bildung* is the traditional German expression that can hardly be translated. The perhaps best translation might be the modern word *literacy* when this is meant in the wider sense.

simply been interpreted as the capacity to apply mathematics in nonmathematical situations. This formulation, however, is the official consensus of all OECD-countries. If it is taken seriously, mathematics classrooms all over the world will be asked to initiate not only high-level problem-solving processes, but also reflections about mathematics itself and its role in the world (cf. Jablonka 2003 for a detailed discussion of this perspective). This does not just include reflections about chances and limits for mathematical models (like in Example 1 on a very simple level), but also includes considerations about the ontological nature of mathematical objects (like in Example 2) or the advantages and difficulties of mechanizations in mathematics (like in Example 3) and many other aspects.

Unlike applying mathematics in nonmathematical situations, which is nowadays widely considered to be an integral part of mathematics itself, many mathematics teachers and researchers in mathematics education consider the claim for the "capacity to identify and understand the role that mathematics plays in the world" as an *additional* task for mathematics education—i.e., a demand that lies *beside* the "pure learning" of mathematics itself. Once the two issues are conceptually separated, and borders between both are defined, questions like the following are naturally asked: "Isn't it necessary to learn mathematics first before reflecting on its role in the world?" or "How should we dedicate any time for these philosophical issues as long as our students are not even able to solve the equations?"

This is why Fischer concluded in 1982 that the primary aim for including philosophical considerations into mathematics classrooms should be to transcend the border between mathematics and other issues around mathematics:

> "Finally, my main aim is the transcending of the (arbitrarily drawn) borders of mathematics. You can even demand to abolish the borders and understand mathematics in a wider sense. In this way, you can reach a didactically fruitful understanding of mathematics as a processual phenomenon that includes applications as well as historical, psychological, sociological and all other imaginable perspectives." (Fischer 1982, 201, my translation)

This claim for a wider understanding of mathematics is the exact central idea of a philosophical program that has been pursued in Darmstadt for twenty five years under the title *General Mathematics* (e.g., Wille 2001), and it has heavily influenced this paper. The program of *General Science* and especially *General Mathematics* starts with the idea that every scientific

discipline should be in critical communication with the general public. Therefore, Rudolf Wille has characterized *General Mathematics* as:

- "the attitude to open mathematics to the public, and to make it principally learnable and criticizable
- the presentation of mathematical developments embedded in its senses, meanings, and conditions
- the teaching of mathematics in its everyday contexts transcending the borders of disciplines
- the discourse about aims, techniques, values, and claims for validity of mathematics"

(Wille 2001, 7, my translation).

General Mathematics is explicitly not considered to be a separate discipline, but an integral part of mathematics. Originally formulated as a program for the scientific disciplines, it is also a suitable and fruitful perspective for describing mathematics for mathematical literacy:

> "Students should learn mathematics primarily as General Mathematics, that especially includes the capacity to understand motives, consequences, purposes, aims, meanings, interpretations, relationships, connections, patterns of thinking and mathematising, analyses, historical backgrounds, typical obstacles and the rich means for expression in mathematics. This does not prevent from learning specific contents or skills. In contrast, it will increase the success for those contents that contribute to the development of the mentioned capacities." (Wille 1995, 54, my translation)

5. CONCLUDING REMARKS

It is exactly this idea of learning mathematics in the broad understanding of *General Mathematics* that led me to answer the initial question "Is there room for a philosophy of mathematics in school practice?" with a "Yes, absolutely!" If we take the normative orientation of mathematical literacy as formulated in the PISA-framework seriously, we cannot exclude philosophical reflections from mathematics classrooms, and the experiences show that there exist strategies and starting points for including them.

Most obviously, philosophical reflections can only take place in an adequate classroom culture. Philosophical reflections will best take place in a community of reflectiveness, dialogue, and mutual respect for diverging thoughts (Prediger 2005). Although still in the minority, there are an increasing number of German teachers who have succeed in moving their

classroom cultures in this direction. This movement makes me optimistic that the situation will change in another thirty years.

To facilitate the dissemination of these ideas, mathematics educators and teacher educators will have to care for the most important obstacles - the teachers' beliefs . Teacher education should

- work on the beliefs of learning and should show by many examples that philosophical reflections are not necessarily difficult and can enhance the learning process before it is completed
- work on the beliefs in mathematics and help develop a broad understanding of mathematics that includes its meanings, senses, and purposes
- work on the beliefs in the aims of mathematics education and help to let the PISA-definition of mathematical literacy become practice

REFERENCES

Becker, Jerry P.; Shimada, Shigeru (1997) (Eds.) *The open-ended approach: a new proposal for teaching mathematics*, NCTM, Reston.

Cobb, Paul; Bouli, A.; McClain, K. and Whitenack, J. (1997). Reflective discourse and collective reflection, *Journal for Research in Mathematics Education,* 28v (39), 258-277.

Davis, Philip J.; Hersh, Reuben (1980). The mathematical experience, Birkhäuser, Boston.

de Lange, Jan (1996). Using and Applying Mathematics in Education, in: Bishop, Alan J. et al. (Eds.) *International Handbook of Mathematics Education*, Kluwer, Dordrecht, 49-97.

de Saint-Exupéry, Antoine (1943). *The little prince*, Egmont, London, 2002 (first published in 1943).

Fischer, Roland (1982). Einige Ansätze zur Philosophie im Mathematikunterricht [Some approaches to philosophy in mathematics classrooms], in: Hans-Georg Steiner (Ed.). *Mathematik – Philosophie – Bildung*, Aulis, Köln, 198-209.

Fischer, Roland (2001). Höhere Allgemeinbildung [Higher general education], in: Fischer-Buck, Anne et al. (Eds.) *Situation – Ursprung der Bildung*, Franz-Fischer-Jahrbuch für Philosophie und Pädagogik 6, Universitätsverlag, Leipzig, 151-161.

Fischer, Roland; Malle, Günther (1985). Mensch und Mathematik. *Eine Einführung in didaktisches Denken und Handeln*, [Mathematics and humans. An introduction into didactical thinking and acting], BI Wissenschaftsverlag, Mannheim, Wien.

Flavell, John H. (1979). Metacognition and cognitive monitoring: A new area of cognitive-developmental inquiry. *American Psychologist*, 34, 906-911.

François, Karen; Van Bendegem, Jean Paul (2004) 'Philosophy of mathematics in the curriculum of mathematics. Questions and Problems Raised by a case study of secondary education in Flanders, Belgium', *Contribution for the 10th International Congress on Mathematics Education, Discussion Group 4*, http://www.icme-organisers.dk/dg04/

Hersh, Reuben (1997). *What is Mathematics, really?* Jonathan Cape, London.

Jablonka, Eva (2003). Mathematical Literacy, in: Bishop, Alan J.; Keitel, Christine; Kilpatrick, Jeremy; Leung, F.K.S. (Eds.) *Second International Handbook of Mathematics Education*, Kluwer, Dordrecht, 75-102.

Kaiser, Gabriele; Maaß, Katja (2007). Modelling in lower secondary mathematics classrooms – problems and opportunities, in: Blum, Werner; Galbraith, Peter L.; Henn, Hans-Wolfgang; and Niss, Mogens (Eds.) *Modelling and Application in Mathematics Education. The 14th ICMI Study*. New York, Springer, 99-108.

Kaune, Christa (2006). Reflection and Metacognition in Mathematics Education - Tools for the Improvement of Teaching Quality. *Zentralblatt für Didaktik der Mathematik*, Heft 38 (4), 350-360.

Keitel, Christine; Kotzmann, Ernst; Skovsmose, Ole (1993). Beyond the tunnel vision. Analysing the relationship of mathematics, technology and society, in: Keitel, Christine; Ruthven, Kenneth (Eds.) *Learning from computers. Mathematics education and technology*, Springer, Berlin, 243-279.

Kitcher, Philip (1984). *The nature of mathematical knowledge*, Oxford University Press, New York.

Krämer, Sybille (1988). *Symbolische Maschinen. Die Idee der Formalisierung im geschichtlichen Abriß*, [Symbolic machines. The idea of formalisation in a historical survey], Wissenschaftliche Buchgesellschaft, Darmstadt.

Krainer, Konrad (1993). Powerful tasks: a contribution to a high level of acting and reflecting in mathematics instruction, *Educational Studies in Mathematics*, 24 (1), 65-93.

Lengnink, Katja (2002). Mathematisches in der Kommunikation [Mathematics in communication], in: Prediger, Susanne; Siebel, Franziska; Lengnink, Katja (Eds.) *Mathematik und Kommunikation, Verlag Allgemeine Wissenschaft*, Mühltal, 121-136.

Lengnink, Katja; Siebel, Franziska (2004) (Eds.) *Mathematik präsentieren, reflektieren, beurteilen*, [Presenting, reflecting and evaluating mathematics], Verlag Allgemeine Wissenschaft, Mühltal.

National Council of Teachers of Mathematics (2000). *Principles and Standards for School Mathematics*, available online under http://standards.nctm.org/

Neubrand, Michael (2000). Reflecting as a Didaktik Construction: Speaking About Mathematics in the Mathematics Classroom, in: Westbury, Ian; Hopmann, Stefan; Riquarts, Kurt (Eds.) *Teaching as a reflective practice. The German Didaktik Tradition*, Lawrence Erlbaum Publishers, London, 251-267.

OECD (1999). *Measuring student knowledge and skills. A new framework for assessment, Organization for Economic Co-operation and Development*, Paris.

Otte, Michael (1977) (Ed.). Philosophie im Mathematikunterricht [Philosophy in mathematics classrooms], *Zentralblatt für Didaktik der Mathematik*, (9) 1 and 2.

Pólya, George (1945). How to solve it, Princeton University Press, Princeton.

Prediger, Susanne (2004). *Mathematiklernen in interkultureller Perspektive. Mathematik-philosophische, deskriptive und präskriptive Betrachtungen*, [Mathematics learning in an intercultural perspective. Philosophical, descriptive and prescriptive considerations], Klagenfurter Beiträge zur Didaktik der Mathematik, Bd. 6, Profil Verlag, München/Wien.

Prediger, Susanne (2005). Developing reflectiveness in mathematics classrooms – An aim to be reached in several ways, *Zentralblatt für Didaktik der Mathematik*, 37 (3), 250-257.

Restivo, Sal; Van Bendegem, Jean Paul; Fischer, Roland (1993) (Eds.) *Math Worlds. Philosophical and Social Studies of Mathematics and Mathematics Education*, State University of New York Press, New York.

Schoenfeld, Alan H. (1992). Learning to think mathematically: Problem solving, metacognition, and sense making in mathematics, in: Grouws, Douglas A. (Ed.). *Handbook of research on mathematics teaching and learning*, Erlbaum, Hillsdale, 287-304.

Sjuts, Johann (2002). Metacognition in mathematics lessons, in: Weigand, Hans-Georg et al. (Eds.) *Developments in mathematics education in German-speaking countries. Selected*

papers of the Annual Conference on Didactics of Mathematics 1999, Franzbecker, Hildesheim, 76-87.

Steiner, Hans-Georg (1989). Philosophische und epistemologische Aspekte der Mathematik und ihr Einfluss auf den Mathematikunterricht, *Mathematische Semesterberichte*, 36 (1), 47-60. (The english version Steiner 1987 does not contain the quoted passage).

Steiner, Hans Georg (1987). Philosophical and epistemological aspects of mathematics and their interaction with theory and practice in mathematics education, *For the learning of mathematics*, 7(1), 7-13.

Thompson, Alba G. (1984) The Relationship Between Teachers Conceptions of Mathematics and Mathematics Teaching to Instructional Practice, *Educational Studies in Mathematics*, 15, 105-127.

Tymoczko, Thomas (1985). *New Directions in the Philosophy of Mathematics*, Birkhäuser, Boston.

von Hentig, Hartmut (1980). *Die Krise des Abiturs und eine Alternative* [The crises of secondary schooling and an alternative], Klett, Stuttgart.

Wille, Rudolf (1995). Allgemeine Mathematik als Bildungskonzept für die Schule [General Mathematics as a guiding idea for mathematical literacy], in: Biehler, Rolf; Heymann, Hans Werner; Winkelmann, Bernard (Eds.) *Mathematik allgemeinbildend unterrichten*, Aulis, Köln, 41-55.

Wille, Rudolf (2001). Allgemeine Mathematik - Mathematik für die Allgemeinheit [General Mathematics. Mathematics for General Public], in: Lengnink, Katja et al. (Eds.) *Mathematik und Mensch. Sichtweisen der Allgemeinen Mathematik, Verlag Allgemeine Wissenschaft*, Mühltal, 3-20.

INTERLUDE 3

Following the previous contribution, which can serve as a direct manual for whoever has direct contact with mathematics education (be it in mathematics teacher training or as a teacher), the next contribution expands on good practices. Whereas the philosophical question in the preceding contribution focused on *what a philosophical reflection precisely is*, the question in the next contribution is broadened to *what mathematics actually is*. Going back to Hersh (1979), the proposition is substantiated that a person's notion of what mathematics is affects the way he/she teaches mathematics. In this contribution, the two aspects of the critical philosophy of mathematics touch. A first point is the philosophy of mathematics itself. What are we talking about when talking about mathematics? A second point is the rendition of, and the inextricable bond between, this view and the way in which mathematical knowledge is passed on to next generations. Thus, this contribution again deals with the values underlying a mathematics curriculum. A significant contribution in this context is the stress put on the values and ideas held by math teachers. They play a very important role at this level, too—in the teaching style, in the way that mathematical themes are introduced, in the way one stresses some things and not others, and in terms of whether one is open to new and alternative forms of mathematical knowledge. Taking this aspect into account, it is important that a future mathematics teacher's attitude is trained to engage in philosophical reflections. A teacher will need to question time and again what values he or she holds so as to be conscious about how these ideas influence his/her practice. In this respect, this development of a teacher's critical mathematical-philosophical consciousness is a necessary condition to raising this selfsame consciousness in pupils. How this can, in fact, be integrated into a mathematics teacher training is a matter that this contribution amply demonstrates through practice at the Faculty of Education, Aristotle University of Thessaloniki, Greece.

INTEGRATING THE PHILOSOPHY OF MATHEMATICS IN TEACHER TRAINING COURSES

A Greek Case as an Example

Dimitris Chassapis
Faculty of Education, Aristotle University of Thessaloniki, Greece

Abstract: The philosophy of mathematics may be assumed to provide a unifying framework that potentially supports an epistemological clarification of mathematical knowledge, as well as a critical reflection on the beliefs and values about mathematical knowledge that a teacher holds in connection with the content and the prevailing practices of mathematics teaching. Thus, the philosophy of mathematics may be considered an essential component of teachers' professional knowledge. In such a perspective, a relevant venture integrating philosophy of mathematics themes in a course of learning and teaching primary school mathematics that is offered to teachers as part of an in-service training program in Greece is presented. The rationale of the venture is outlined, selected examples are briefly presented, and issues that arise in its implementation are reported.

Key words: Teacher training, didactics, teachers' beliefs , primary school, nature of mathematics

1. BACKGROUND ORIENTATION

Arguments from three different, although interrelated, perspectives that are briefly presented in the following support directly or indirectly the thesis that the philosophy of mathematics has to be considered an indispensable component of teachers' professional knowledge. This thesis constitutes the starting point of a venture attempting to integrate selected themes from the

philosophy of mathematics into teachers' training course on primary school mathematics, which is reported in this paper.

The first argument asserts the direct association of a philosophy of mathematics with fundamental features of mathematics education. Many years ago, Thom claimed that a philosophy of mathematics has powerful implications for educational practice pointing out that "In fact, whether one wishes it or not, all mathematical pedagogy, even if scarcely coherent, rests on a philosophy of mathematics" (Thom 1973, 204). Hersh, few years later, emphasised this claim noting that

> "One's conception of what mathematics is affects one's conception of how it should be presented. One's manner of presenting it is an indication of what one believes to be most essential in it. ... The issue, then, is not, What is the best way to teach? but, What is mathematics really about? ..." (Hersh 1979, 33).

Generalizing and exemplifying further this position of association between a philosophy and didactics of mathematics, Steiner notes that

> "Concepts for teaching and learning mathematics—more specifically: goals and objectives (taxonomies), syllabi, textbooks, curricula, teaching methodologies, didactical principles, learning theories, mathematics education research design (models, paradigms, theories, etc.), but likewise teachers' conceptions of mathematics and mathematics teaching as well as students' perception of mathematics—carry with them or even rest upon (often in an implicit way) particular philosophical and epistemological views of mathematics" (Steiner 1987, 8).

The argument of association of a philosophy of mathematics with fundamental features of mathematics education, exemplified in the above quotations, has been principally developed on the grounds of theoretical analyses.

A second argument is that teachers' ideas, views, conceptions, or beliefs (the terms are dependent on conceptual frameworks and thus on adopted theoretical perspectives) about mathematics, its learning, and teaching, implicitly reflect, or are related to, a philosophy of mathematics. These teachers' beliefs about mathematics, in their turn, play a significant role in shaping characteristic patterns of their didactic practices, as it is widely accepted nowadays, either on the basis of empirical evidence (Thompson 1984, 1992) or on a philosophical realm (Lerman 1983, 1990; Steiner, 1987; Ernest, 1989).

Teachers' beliefs about mathematics are personally held mental constructs about the nature of mathematical knowledge, which includes,

among others, beliefs about the origins of mathematical knowledge, the nature of mathematical knowledge as a discipline, the nature of mathematical problems and tasks, the relationships between mathematical knowledge and empirical reality, and, in particular, about the applicability and utility of mathematical knowledge, the nature of mathematical knowledge as a subject taught in schools (Törner 1996), as well as beliefs about oneself as a learner and user of mathematics, and more generally, beliefs about the process of learning mathematics (Ernest 1989, Pehkonen 1994).

Teachers' beliefs about mathematics reflect, or are related to, a philosophy of mathematics and, in fact, constitute a kind of *practical philosophy of mathematics*, which, as a complex, practically-oriented set of understandings, regulates and shapes to a great extent the teachers' thoughts and practices within classrooms, although subject to the constraints and contingencies of the school context. Moreover, this teachers' *practical philosophy of mathematics*, often takes precedence over knowledge, shaping the interpretation of their held knowledge and selectively admitting or rejecting new knowledge. The process of how, why, when, and under what circumstances beliefs about mathematics are adopted and defined by the individual teacher is not yet clearly and definitely described by relevant literature. In any case, their discussion falls beyond the aim of this paper.

Therefore, assuming an association of a philosophy of mathematics with mathematics education issues on the one hand and teachers' philosophical and epistemological views of mathematics on the other, then philosophy of mathematics per se has to be considered as an indispensable component of teachers' professional training in mathematics learning and teaching—a training, however, that aims at enabling teachers to develop a questioning stance towards dominant canons of mathematics education, a critical reflection of their personal didactical practices, and an increase in their professional autonomy in teaching mathematics. Or, more broadly, it needs to be a training that aims to support teachers in becoming reflective practitioners, playing an important role in the definition of the purposes and goals of their work, as well as on the means to attain them, and therefore, participating in the production of knowledge about teaching mathematics. It must be a knowledge about teaching mathematics, however, with an element of critique of the established standards.

The third argument—the assumption that a philosophy of mathematics is directly associated to a deeper understanding of mathematics as the subject matter knowledge of teaching it—seems reasonably hard to doubt.

In recent years, the importance of mathematical knowledge has been well documented in the literature, and the lack of it has been linked to less competent mathematics teaching (Rowland, Martyn, Barber and Heal, 2000,

2001) and over-reliance on commercial schemes (Millett and Johnson 1996). It has been well documented, as well, that understanding mathematics for teaching entails both knowledge *of* mathematics and knowledge *about* mathematics or, using Shulman's distinction, *substantive* and *syntactic* knowledge of mathematics (Shulman 1987, 8). Knowledge *of* mathematics includes both propositional and procedural knowledge (i.e., concepts, principles, facts, and the ways that they are organized). Ball argued that to teach mathematics effectively, teachers must have knowledge of mathematics, characterized by an explicit conceptual understanding of the principles and meanings underlying mathematical procedures and by connectedness—in contrast to compartmentalization of mathematical topics, rules, and definitions (Ball 1990). Knowledge *about* mathematics includes an understanding of the nature of mathematical knowledge and the mechanisms through which new knowledge is introduced and accepted in the community of mathematicians, as well as knowledge about proofs, rules of evidence, and structures (Schwab 1978).

In summary, the philosophy of mathematics, as well as the history and sociology of mathematics, are essential components of the domain of knowledge about mathematics. Against this background, a venture of integrating selected themes from the philosophy of mathematics into teacher training has been undertaken over the last five years as a constituent element of a course on learning and teaching primary school mathematics. This course is offered to primary school teachers by the author of this paper, as part of an in-service training program run by the Primary Education Department of the Aristotle University of Thessaloniki, Greece.

In accordance with the aims of the course, the philosophy of mathematics is conceived in a descriptive and social perspective and is considered to account for:

- Mathematical knowledge: its nature, justification, and genesis
- The objects of mathematics: their nature and origins
- The application of mathematics: its effectiveness in science, technology, and other realms
- Mathematical practice: the activities of mathematicians, both in the present and the past (Ernest 1991, 27).

In the following section, main features of the rationale of this venture are outlined, and issues that have arisen by an overall evaluation of the course are briefly reported and commented upon.

2. INTEGRATING THEMES FROM PHILOSOPHY OF MATHEMATICS IN A TEACHER TRAINING COURSE: A GREEK PROJECT

The philosophy of mathematics may be considered an essential component of teachers' professional knowledge. In such a perspective, a relevant venture integrating philosophy of mathematics themes in a course of learning and teaching primary school mathematics that is offered to teachers as part of an in-service training program in Greece is presented.

2.1 The Rationale and its Background

Any introduction of the philosophy of mathematics to teacher' training courses has to be designed on a basis framed by the proposed answers to the following core questions:

- *Content*: What topics from the philosophy of mathematics should be taught to teachers?
- *Method*: What methods are most appropriate to teach them to teachers?
- *Incorporation:* What relationships should be established between courses of philosophy of mathematics, mathematics, and didactics of mathematics offered to teachers in a training program?

Answering these questions first, and in accordance with the aims outlined above, a course was offered including topics from the philosophy of mathematics, adopting mainly an informative approach, designed and implemented on an experimental basis, and offered to teachers as a supplementary course to the main course on learning and teaching primary school mathematics. Both courses were one semester long, each taught in one-and-a-half hour lecture followed by a one-and-a-half hour discussion session. After two semesters of implementation, however, this option of distinct, although supplementary, courses was evaluated and found to have a non-significant impact on teachers' thinking. The main reason for failure was ascribed to a revealed inability of teachers to perceive any relevance between the topics of philosophy of mathematics and their immediate teaching interests.

In this account, two prerequisites of any attempt to introduce the philosophy of mathematics in a teacher training course were noticed. First, an apparent relevance of topics between the philosophy of mathematics and mathematics teaching may be a crucial factor for teachers' motivation. Second, thought-provoking questions regarding mathematics taught in

primary schools could function as a catalyst in attracting teachers' interest and involvement in the philosophy of mathematics. At the same time, an empirical investigation showed that primary school teachers' prevailing beliefs about mathematics were dominated by a conception of mathematics as a fixed, predictable, absolute, certain, value-free, culture-free, and applicable body of knowledge involving a set of facts, rules, and procedures to be used in the pursuance of some external end (Chassapis 2003). In addition, teachers' poor mathematical backgrounds and a rather narrow view of mathematics as an academic discipline were also evidenced by relevant studies and ought to be considered.

Taking into account the previously summarized issues, and in addition to the main aims of teachers' training, the scheme of integrating themes from the philosophy of mathematics into the main course on learning and teaching primary school mathematics has been developed and implemented using the following rationale:

- Themes from the philosophy of mathematics have been selected and were developed along three threads stemming from the primary mathematics content:
    - Concepts (cardinal and ordinal number concepts, definitions of natural and rational numbers, questions on the nature and properties of numbers, irrational numbers, numeration systems as cultural constructs, continuity and infinity issues arisen from rational and real number concepts)
    - Processes (definition, justification and proving in mathematics, the what and the why of axiomatic systems in mathematics)
    - Applications (problem-solving and relationships between mathematical knowledge and empirical reality)
- Themes from the philosophy of mathematics are introduced and discussed using thought-provoking questions, which create dissonant situations for the teachers and thus motivate them to be actively involved in discussion, learning, and reflective thinking
- Questions concerning the nature of mathematical knowledge and practice have arisen recurrently, on many occasions, during lectures and discussions of the course—for instance, issues concerning the use of manipulatives in teaching particular mathematical concepts

The organizing concepts of this rationale are *themes* and *thought-provoking questions*. By *themes* in the philosophy of mathematics we mean

collections of learning experiences that assist students in relating their learning to questions that are important and meaningful for them, as well as practice-bounded (Freeman and Sokoloff, 1996). Themes are the organizers of the philosophy of mathematics content, which is presented and discussed using questions meaningful to the teachers as starting points. They are intended to give meaning and direction to the reflection and learning process (Perfetti and Goldman, 1975). Finally, the thematic approach seems to provide an environment where knowledge can be individually and socially constructed, so it may be considered to be associated with constructivist ideas of knowing.

By *thought provoking-questions*, which create dissonant situations for the teachers and thus motivate them to be actively involved in approaching the philosophy of mathematics, we mean questions create perplexity, challenge beliefs, spot questionable issues, and potentially foster a conceptual reconstruction of mathematical knowledge and pedagogy for the teachers.

2.2 The Content and Organization of the Course

In the outline of the course on learning and teaching primary school mathematics that follows, a record is found of the attempted articulation of thematic units from the philosophy of mathematics into the content of this course. Besides the themes from the philosophy of mathematics, selected issues from the historical development of mathematical concepts are occasionally introduced to emphasize that mathematics is a constantly changing creation of human activity and not a fixed and finished, a priori existing product, which one is expected to discover. Inset examples (see Appendix at the end of this chapter) indicate the types of questions introducing the philosophy of mathematics issues, which are posed during the lectures of the course. The questions are put forward in various wordings, depending on the case at hand.

The following is an outline of the course on learning and teaching primary school mathematics, integrating themes from the philosophy of mathematics. Items in *italic* represent issues of philosophy.

UNIT 1 Mathematical concepts and their characteristics

Topics

- Fundamental features of the mathematical concepts and the initial difficulties children find in comprehending them
- The articulation and organization of the mathematical concepts in systems

Questions / Thematic unit from philosophy of mathematics
Why the organization of mathematics concepts in axiomatic systems?
Euclid: the axiomatization of geometry
Peano: the axiomatization of natural numbers
Gödel: the incompleteness theorems
What does the axiomatic organization of mathematics concepts mean for the learning of mathematics?
Further discussion issues: definition, justification, and proving in mathematics (example: Pythagorean theorem proofs)

- The graphical and symbolic representations and the linguistic expressions of the mathematical concepts
- The relationships of mathematical concepts to empirical reality

Questions / Thematic unit from philosophy of mathematics
Are mathematical concepts and truths discovered or invented?
Realism and anti-realism in mathematics

UNIT 2 Classes, sets and relations in mathematics

Topics

- Classes, sets and relations
- Equivalence and ordering relations

Questions / Thematic unit from philosophy of mathematics
Are the collections of objects used in primary mathematics classes or sets?
And how is the difference between classes and sets conceived?
The Russell's paradox and its consequences
Finite and infinite sets
Whole and part
The Cantor's paradox
The continuum hypothesis
The question of foundations of mathematics: answers, approaches and schools of thought (logicism, formalism and intuitionism)

- Logical operations
- The formation and development of logical-mathematical concepts and relations in childhood: Piagetian and socio-cultural perspectives

UNIT 3 The number concepts

Topics

- The concept of cardinal and ordinal number

  Questions / Thematic unit from philosophy of mathematics
  What does the numerocity of a class of objects mean?
  Using potency of classes of objects, which concept of cardinal and ordinal numbers is implicitly constructed in primary mathematics?
  What are numbers: objects or properties?
  What is the relation of the so constructed (and implicitly defined) concept of cardinal and ordinal numbers to counting and its outcomes?
  Number definition approaches and issues in philosophy of mathematics (Pythagoreans, Cantor, Peano, Frege)
  Do different number definitions impact any differences in teaching number concept? (Example: Cardinal and ordinal numbers in primary mathematics)

- Numeration systems and linguistic numerical expressions

  Questions / Thematic unit from philosophy of mathematics
  Why are various numeration systems invented and used over time and across cultures?
  Numeration systems as cultural constructs
  The cultural aspects of mathematical activity
  Piagetian and constructivist approaches to the formation of number concepts

- Socio-cultural approaches to the acquisition of number concepts

UNIT 4 Number operations

Topics

- Mathematical definitions and properties of number operations

  Questions / Thematic unit from philosophy of mathematics
  How is number addition and multiplication defined in mathematics?
  Do different definitions of number operations impact any differences in their teaching?
  A further discussion issue: Definition in mathematics

- Number operations in the decimal numeration system—Algorithms
- The mapping of number operations to real world situations

Questions / Thematic unit from philosophy of mathematics
What truth is expressed by a number operation and, more generally, by a mathematical statement?
Is a truth about objects, concepts, or neither?
The question of mathematical truth
The indispensability of mathematics arguments

UNIT 5 Expansions of number concept

Topics

- On the mapping of numerical concepts to real world situations: Counting, ordering and measuring
- Natural numbers and integers
- The concept of zero in natural numbers and in integers
- The fraction concept and the rational numbers
- Decimals and percentages

Questions / Thematic unit from philosophy of mathematics
What happens when ordering rational numbers?
Discreteness and continuity
Infinity and its paradoxes: From Zeno to Cantor and beyond

- The concept of irrational numbers
- The concept of real numbers

Questions / Thematic unit from philosophy of mathematics
Is any number modeled to an attribute of empirical reality and vice-versa?
Does square root of 2 exist?
The existence of mathematical entities
In what sense, if any, do mathematical entities (numbers, for instance) exist?
The nature of mathematics concepts: Mathematical entities and empirical objects
Approaches and schools of thought on the nature of mathematics and mathematical activity

UNIT 6 On methods and media for teaching mathematics in primary classroom

Topics

- Methods for teaching mathematics in primary classroom
- Tools and materials for teaching mathematics in primary classroom

Question / Thematic unit from philosophy of mathematics
Is there any, and what, meaning in the discovery methods for teaching mathematics concepts?
Are mathematics concepts embodied in manipulatives?
(Example: Multi-base Arithmetic Blocks)
The ontology of mathematics and its implications for mathematics education
A further discussion issue: What is mathematics after all?

Schematically, and in a figurative language, the attempted mode of articulation of philosophy of mathematics themes into the units of the course on learning and teaching primary school mathematics could be described as a spiraling one.

2.3 Feedback and Evaluation of the Course

The course, as outlined above, has been offered for the last two years. Each semester, after the completion of the course, the teachers that have participated are asked to write a short anonymous report evaluating the main aspects of the course, describing the most important gains they enjoyed from attending the course and spotting the main obstacles that they encountered in following the lectures and participating in the discussions of the course. All reports are being carefully and read many times, so the evaluative comments made by the teachers can be clearly elicited from their writing and the derived data recorded and analyzed.

From the analysis of the evidence collected over the two years of this course's implementation, a number of conclusions were drawn. Most of the teachers point out that their main gain from attending the course was incitement for conceptual change, or an actual conceptual reconstruction they engaged in. In many cases, it was vaguely described as a *change of mind* about mathematics. In every case, this re-conceptualization of the nature of mathematics led them to view the discipline from a perspective different from their own long-held perspective. This perspective on mathematics that is *new* for them has, among its impacts, the adoption of a critical approach towards many dominant standards of, and prevailing

practices in, mathematics education. This critical approach is mostly emanated, according to all available evidence, from the possibility of questioning what is taken for granted in school mathematics—offered by their involvement in the philosophy of mathematics.

The themes from the philosophy of mathematics that are obviously appreciated by the teachers, and which seem to have the greatest effect on their thinking, are those which offer explicit opportunities for challenging their conceptions about teaching particular concepts of school mathematics (e.g., numbers) or elements of their teaching models (e.g., the use of manipulatives as embodiments of mathematical concepts or processes).

The most serious problem encountered during the implementation of the course originated from the poor mathematical backgrounds of the teachers. The lack of specific mathematical knowledge does not permit them to comprehend specific questions and ideas from the philosophy of mathematics, and delimits the extent, depth, and quality of the relevant discussions. Infinity and continuity issues are the most characteristic examples. Most of the teachers enrolled in this and other in-service courses, had already graduated from Higher Teachers' Training Colleges offering three-year courses (which was operative in Greece up to 1985, when they were replaced by university departments), so their original mathematical background is relatively poor. The outcomes of this course, as far the philosophy of mathematics is concerned, are crucially dependent on teachers' mathematical backgrounds. The repeated conclusion of many research studies, that teachers' existing knowledge and beliefs are critical in shaping what and how they learn from teacher training experiences, seems to be validated once more in our case.

A second problem, crucial for the efficacy of the course, appeared to be the lack of appropriate resources in Greek language on the philosophy of mathematics intended for primary school teachers to meet their reading requirements, which was necessary for their constructive participation in the course learning activities.

A final issue—not induced from teachers reports but from the author's experience—which must be registered in the assessment of the course, is the demand in time and work both for the preparation and for the running of each course session, placed, as a rule, on the professor.

3. CONCLUDING COMMENTS

Until recently, mathematics education has been developed in a way that is scarcely related to the philosophy of mathematics. Rationales and practical proposals in mathematics education are even now mainly informed by

educational and psychological research, disregarding ideas coming from disciplines that study the nature of mathematics, as do the philosophy, history, and sociology of mathematics. Nowadays, however, this situation seems to be changing through the work of many scholars and teacher trainers.

In this paper, I have presented an attempt to integrate themes from the philosophy of mathematics in a teachers training course on learning and teaching mathematics in primary school, aimed to contribute into supporting teachers in becoming reflective practitioners. The overwhelmingly positive feedback from the majority of teachers that attended this course, was an initially surprising finding. It must be recognized, however, that no single course is a panacea for promoting aims concerning teachers' reflective thinking and acting. Teacher courses, even sequences of courses over multiple semesters, seem to make some difference, but may be insufficient to promote the kinds of changes in the conceptual understanding and ideology of teachers necessary to achieve outcomes compatible with the ideals of reflective practice. It may be that to fight against those years of learning reinforces the status quo, and that the slow accumulation and layering of reconstructive experiences over many other years will be required.

Finally, there is no doubt that a careful review of the course syllabus, together with empirical evidence or experiences from similar efforts, will likely reveal additions and deletions or a reorganization of the course content. Any such review is welcomed and necessary to establish a dialogue for optimal approaches to the integration of the philosophy of mathematics in prospective and in-service teachers training programs.

APPENDIX

Example 1: Pythagorean Theorem Proofs

In right-angled triangles, the square on the side opposite the right angle equals the sum of the squares on the sides containing the right angle.

1.1: Euclid's proof based on deductive reasoning:
(Euclid's *Elements*, Book 1, Proposition 47)

Figure 1: Euclid's Proof of the Pythagorean Theorem.

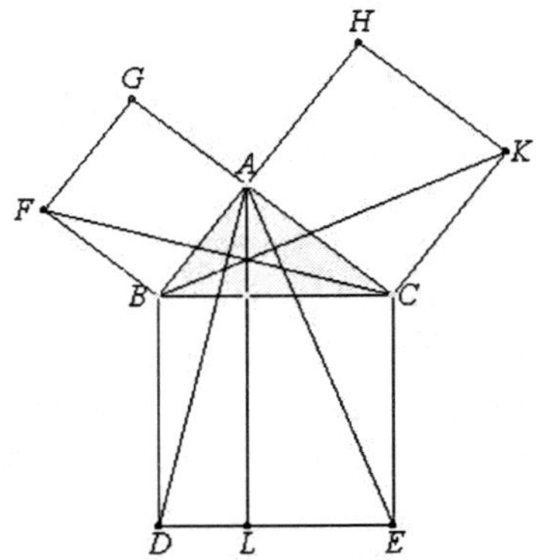

Let ABC be a right-angled triangle having the angle BAC right. The square on BC equals the sum of the squares on BA and AC. Describe the square BDEC on BC, and the squares GB and HC on BA and AC. Draw AL through A parallel to either BD or CE, and join AD and FC. Because each of the angles BAC and BAG are right, it follows that with a straight line BA, and at the point A on it, the two straight lines AC and AG (not lying on the same side) make the adjacent angles equal to two right angles. Therefore, CA is in a straight line with AG. For the same reason, BA is also in a straight line with AH. Because the angle DBC equals the angle FBA (each is right), add the angle ABC to each. The whole angle DBA equals the whole angle FBC. Because DB equals BC, and FB equals BA, the two sides AB and BD equal the two sides FB and BC, respectively, and the angle ABD equals the angle FBC. Therefore, the base AD equals the base FC, and the triangle ABD equals the triangle FBC. Now, the parallelogram BL is double the triangle ABD, for they have the same base BD and are in the same parallels BD and AL. The square GB is double the triangle FBC, for they again have the same base FB and are in the same parallels FB and GC. Therefore, the parallelogram BL also equals the square GB. Similarly, if AE and BK are joined, the parallelogram CL can also be proved equal to the square HC. Therefore, the whole square BDEC equals the sum of the two squares GB and HC. And the square BDEC is described on BC, and the squares GB and HC on BA and AC. Therefore, the square on BC equals the sum of the squares on BA and AC. In right-angled triangles, the square on the side opposite the right angle equals the sum of the squares on the sides containing the right angle.
Quod erat demonstrandum (Q.E.D).

1.2: Liu Hui's proof based on the Chinese dissection process:
(Liu Hui, commentary on the Jiuzhang suanshu, third century B.C. In Swienciki, L.W. *The Ambitious Horse. Ancient Chinese mathematics problems,* Emeryville, CA: Key Curriculum Press, 2001, 58)

Figure 2: Liu Hui's Proof of the Pythagorean Theorem.

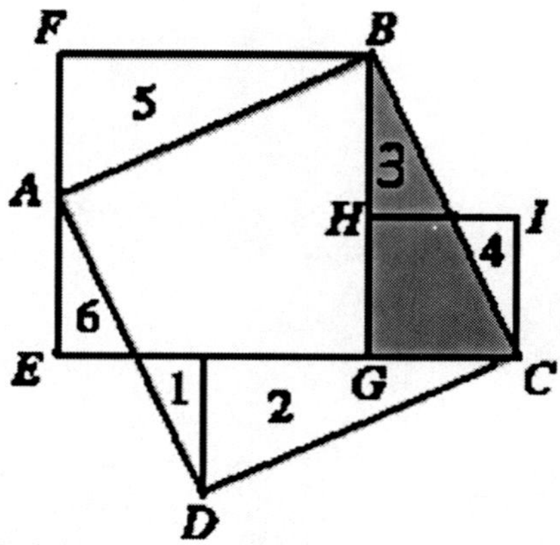

Start with square ABCD, whose side is c and area is c^2. Remove pieces 1, 2, and 3 by translation and rotate to form pieces 4, 5, and 6, respectively. The original square has now been transformed into two new squares: 1) square EFBG of side b and area b^2, 2) square GHIC of side a and area a^2.

We write Area (GHIC) + Area (EFBG) = Area (ABCD) or $a^2+b^2=c^2$.

Questions:

Which proof of the Pythagorean theorem is accepted by the discipline of mathematics, is included in mathematics textbooks, and is used in teaching mathematics?

Why?

Why are hands-on activities not accepted as justification arguments in mathematical proofs?

What is the aim of a proof in mathematics?

Why is deductive, and not creative, thinking mostly valued in mathematics?

Example 2: Cardinal and Ordinal Numbers in Greek Primary Mathematics

Source: *My mathematics*, 1st primary school grade, Athens: OEDB, 2003, 26, 74. (in Greek)

Figure 3: Cardinal and Ordinal Numbers in Greek Primary Mathematics.

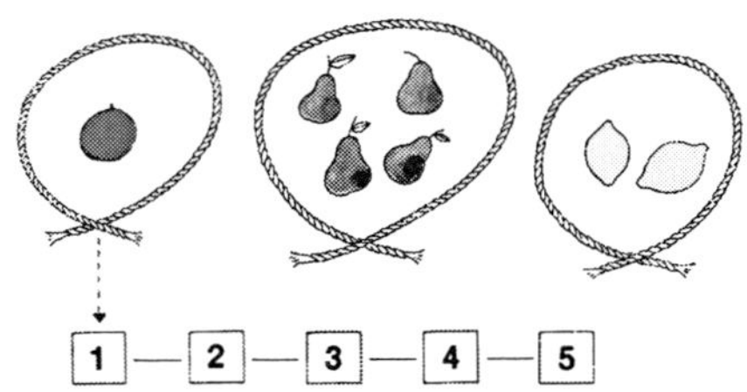

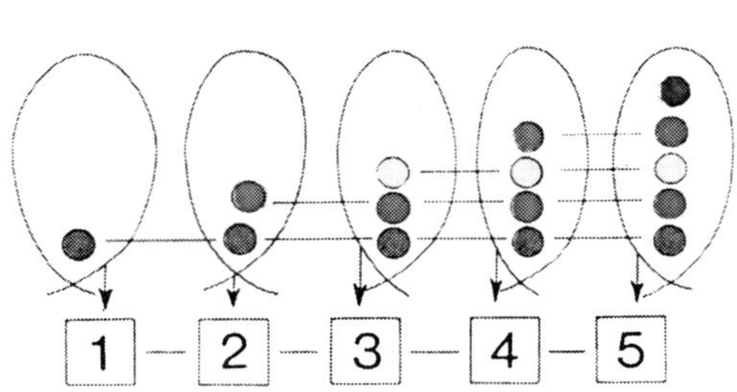

Questions:

Using the potency of classes of objects, which concept of cardinal and ordinal numbers is implicitly defined in primary mathematics?

What are numbers: objects or properties?

Example 3: Multi-base Arithmetic Blocks (Base Ten)

How many blocks do you see?
These blocks stand for a certain number. What number do you think they stand for?

Figure 4: Multi-base Arithmetic Blocks (Base Ten).

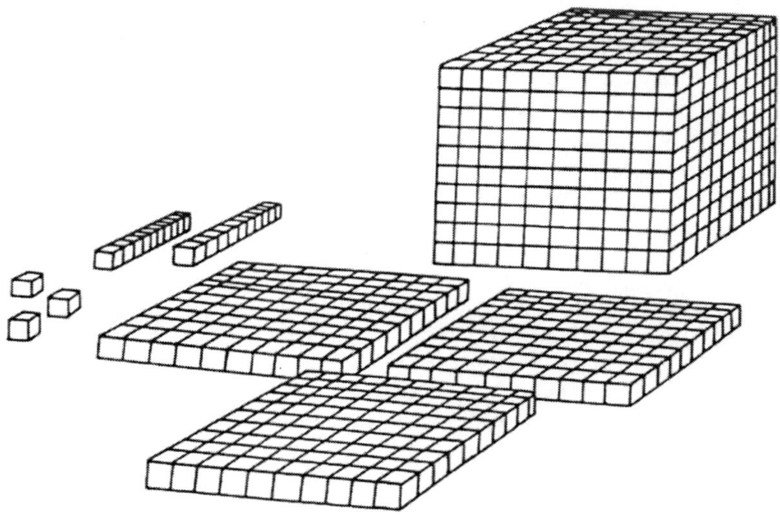

Answers given by 3rd graders:

9 (counting discrete objects, disregarding differences in size and the markings on the blocks)

About 1¼ (measuring using as reference unit 1 (block/cube) + the rest make about a quarter)

923 (measuring surfaces/areas using as reference unit 1 surface = 100 therefore 6 x 100 =600 on the cube + 3x100=300 flat pieces + 20+3)

1.323 (decimal number system)

Questions:
Which interpretation do you consider as "correct"?
Why?

REFERENCES

Ball, D.L. (1990). The mathematics understanding that prospective teachers bring to teachers education, *Elementary School Journal,* 90, 449-466.

Ball, D. L (1991). Research on teaching mathematics: Making subject-matter knowledge part of the equation. In J. Brophy (Ed.), Advances in research on teaching, Vol. 2, Greenwich, CT: JAI Press, 1-48.

Chassapis D. (2003). Greek primary schools teachers' beliefs about mathematical knowledge, *Proceedings of the 3rd Mediterranean Conference on Mathematical Education,* Athens: Hellenic Mathematical Society, 409 – 417.

Ernest, P. (1989). The Impact of Beliefs on the Teaching of Mathematics. In P. Ernest (Ed.), *Mathematics Teaching: The State of the Art*, London: Falmer Press, 249-254.

Ernest, P. (1991). *The philosophy of mathematics education*, London: Falmer.

Freeman, C., and Sokoloff, H.J. (1996). Toward a theory of theory of thematic curricula: Constructing new learning environments for teachers and learners. *Education Policy Analysis Archives*, 3(14), 1-19.

Hersh, R. (1979). Some proposals for revising the philosophy of mathematics. *Advances in Mathematics*, 31 (1), 31-50. Reprinted in T. Tymoczko (Ed.), (1986), New directions in the Philosophy of Mathematics, 9-28, Boston: Brikhauser.

Lerman, S. (1983). Problem solving or knowledge centred: The influence of philosophy on mathematics teaching, *International Journal of Mathematics Education in Science and Technology,* 14 (1), 59–66.

Lerman, S. (1990). Alternative perspectives of the nature of mathematics and their influence on the teaching of mathematics, *British Educational Research Journal* 16(1), 53–61.

Millett, A. and Johnson, D. C (1996). Solving teachers' problems? The role of the commercial mathematics scheme. In D. C. Johnson and A. Millet (Eds.) *Implementing the Mathematics National Curriculum: Policy, politics and practice*. New BERA dialogues series, 1. London: Paul Chapman Publishing Ltd, 54-70.

Pehkonen, E. (1994). On teachers' beliefs and changing mathematics teaching, *Journal für Mathematik-Didaktik,* 15 (3/4), 177-209.

Perfetti, C.A., and Goldman, S.R. (1975). Discourse functions of thematization and topicalization, *Journal of Psycholinguistic Research*, 4(3), 257-271.

Restivo, S., Van Bendegem, J.P. and Fischer, R. (Eds.) (1993). *Math worlds. Philosophical and social studies of mathematics and mathematics education*, State University of New York Press: New York.

Rowland, T., Martyn, S., Barber, P. and Heal, C. (2000). Primary teacher trainees' mathematics subject knowledge and classroom performance. In T. Rowland and C. Morgan (Eds.) *Research in Mathematics Education Volume 2: Papers of the British Society for Research into Learning Mathematics*, London: British Society for Research into Learning Mathematics, 3-18.

Rowland, T., Martyn, S., Barber, P. and Heal, C. (2001). Investigating the mathematics subject matter knowledge of pre-service elementary school teachers. In M. van den Heuvel-Panhuizen (Ed.) *Proceedings of the 23rd Conference of the International Group for the Psychology of Mathematics Education Vol 4,* Utrecht, The Netherlands: Freudenthal Institute, Utrecht University, 121-128.

Schön, D. A. (1983). *The Reflective Practitioner: How Professionals Think in Action.* Aldershot Hants: Avebury.

Schwab, J. J. (1978). Education and the structure of the disciplines. In I. Westbury and N. J. Wilkof (Eds.) *Science, Curriculum and Liberal Education,* Chicago: University of Chicago Press, 229-272.

Shulman, L. S. (1987). Knowledge and teaching: Foundations of the new reform, *Harvard Educational Review*, 57(1), 1-22.

Steiner, H.-G. (1987). Philosophical and epistemological aspects of mathematics and their interaction with theory and practice in mathematics education, *For the Learning of Mathematics* 7(1), 7-13.

Thom, R. (1973). Modern mathematics: does it exist? In A.G. Howson (Ed.), *Developments in Mathematics Education: Proceedings of the Second International Congress on Mathematics Education.* Cambridge: Cambridge University Press, 194-209.

Thompson, A. (1984). The relationship of teachers' conceptions of mathematics teaching to instructional practice, *Educational Studies in Mathematics*, 15, 105-127.

Thompson, A. (1992). Teachers' beliefs and conceptions: A synthesis of the research. In D. A. Grouws (Ed.), *Handbook of research on mathematics teaching and learning,* New York: Macmillan, 127-146.

Törner, G. (1996). Views of German mathematics teachers on mathematics. In Weigand, H.-G. et. al. (Eds.) *Selected Papers from the Annual Conference on Didactics of Mathematics, Regensburg, 1996,* Gesellschaft für Didaktik der Mathematik, 116-129.

INTERLUDE 4

After the examples of ways a philosophy of mathematics can be implemented in mathematics teacher training and in the classroom itself—which gave rise to philosophical questions like "what is a mathematical philosophical reflection?" and particularly "what *is* mathematics, really?" and "what is its nature?", it is time for mathematics history to intervene. It is hard to present an image of what mathematics is without referring to the story of its origins. A history of mathematics is indispensable if we are to adjust the image of mathematics as a discipline, which, like other scientific disciplines, is imbedded socially and culturally. Long and beautiful stories have been written about the history of mathematics, and even these illustrate that there are different ways in which this history is told (Kline 1972, Struik 1948, Restivo 1992). As is the case with other origin stories, there apparently is not just one history of mathematics. Within the great story, we can zoom in on one facet or another. In the next contribution we are taken on an adventure tour of *symbolic algebra.* Even though history in itself is a fascinating story, its surplus value—as far as this book is concerned—lies in the fact that the author indicates what a math teacher can do with such a historical approach. Should you know about the history of the number zero to grasp the most confusing mathematical conventions related to it? It can be done without history, that is a fact. This is shown very commonly in traditional mathematics education, in which a number of short biographies of male mathematicians—fathers of mathematics—are presented, but not considered *material for the exam.* A history of mathematics, then, is little more than a means to embellish a handbook. Clearly, this is not what mathematics history contributes to mathematics education. Mathematics education without history, or a somewhat special interpretation of history, is possible. The argument in the following contribution, however, is that the history of mathematics can stimulate the insight that mathematics is characterized by a plurality of methods. This insight will be elaborated on in a later chapter, where the relevance of the intuitive mathematical method is shown, as well as the educational praxis, and the so dominant deductive method (see Van Moer, this volume). In addition, a history of mathematics can demonstrate the dynamics to which mathematical concepts are subject. This should heighten the philosophical alertness of pupils. Here is yet another contribution touching upon the value system behind the curriculum of mathematics. Here, too, it is illustrated how we can give mathematics a place at a given time, in a given culture and context—a theme which is explored further in this book (see Pinxten and François, this volume).

LEARNING CONCEPTS THROUGH THE HISTORY OF MATHEMATICS

The Case of Symbolic Algebra

Albrecht Heeffer
Centre for Logic and Philosophy of Science - Ghent University, Belgium

Abstract: The adolescent's notion of rationality often encompasses the epistemological view of mathematics as knowledge which offers absolute certainty. Several findings such as Gödel's theorem and the construction of a strict finite arithmetic, however, provide strong arguments against that view. The static and unalterable mode of presentation of concepts in the mathematics curriculum, rather than lack of knowledge of metatheory, contributes to this misconception. I will argue that the conceptual history of mathematics provides excellent opportunities to convey the basic epistemological and ontological questions of the philosophy of mathematics in mathematics education. In particular, the emergence of the concept of an equation will be presented in a historical context. Such examples will alert students of the relativity of mathematical methods, truth, and knowledge, and will put mathematics back in the perspective of time, culture, and context.

Key words: History of algebra, concept formation, absolute truth, inconsistency

1. INTRODUCTION

In this chapter, we argue for the integration of the history of mathematics in mathematics education. Although history is not the same as philosophy, we believe that the history of mathematics provides many opportunities to convey basic philosophical concepts concerning the epistemological and ontological aspects of mathematics.

On the epistemological level, a conceptual history of mathematics raises questions such as:

- How are concepts formed in mathematics?
- Which factors influence or change the meaning of concepts?
- Is there an internal logic and order in the development of mathematical concepts?
- What is the role of symbolism in mathematical knowledge?
- What constitutes a valid proof in mathematics?

Concerning ontology, the history of mathematics provides challenging arguments in the realism-constructivism debate. If the meaning of basic mathematical concepts, such as number or equation, changes during the development of mathematics, what happens to the ontological status of these concepts? We will explore how these questions can be approached within the teaching of elementary algebra.

Several studies have been published on the use of the history of mathematics in mathematics education.[1] Arguing for the use of the history of mathematics may therefore seem to be a superfluous task. The official curriculum for secondary education in Flanders (Belgium) defines the role of the history of mathematics explicitly:[2]

> Mathematics education is necessarily connected with other disciplines. Mathematics itself has developed through centuries in close connection with prevailing opinions and problems. Today, certain historical contexts still provide useful starting points to approach specific mathematical concepts and educational topics. The historical context shall therefore be integrated in our curriculum.

When looking for concrete guidelines of how to integrate the history of mathematics, however, the examples offered by the plan are disappointing. We only find generalities such as "an approach with examples from architecture and painting can illustrate the role of mathematics in the development of certain art forms" (ibid. 11) and "assignments can be given to research historical facts, such as an internet search for mathematicians, important mathematical theorems, mathematical illustrations and applications" (ibid. 28). The history of mathematics is forced into an illustrative role. History delivers the pictures for lighting up dreary textbooks, to force a connection with other disciplines, and to keep students busy between other assignments. An *integrated* view on mathematics

1 Some representative collections are Callinger 1996 and Fauvel and van Maanen 2000.

2 Cited from the education plan of the first two years of secondary school used by Catholic schools (2002, 29, my translation). The education plans of other schools in Belgium and in other European countries employ very similar formulations. For an overview of the place of history in mathematics education in several countries, see the ICMI study of Fauvel and van Maanen 2000, Chapter 1.

education, in which the history of mathematics has a methodological and philosophical relevance, is absent.

We will provide some basic arguments for the integration of the history of mathematics into mathematics education. The first addresses the epistemological status of mathematics. The adolescent's notion of rationality often encompasses the epistemological view of mathematics as knowledge that offers absolute certainty. He probably has heard of a geometry in which the parallel postulate does not hold, but most likely believes that Euclidian geometry is the *real one*. We can assume that he is not familiar with Gödel's theorems and undecidability. It is further unlikely that he has been taught about the existence of inconsistent arithmetic that performs finite calculations as correct as traditional arithmetic. These findings provide strong arguments against the view that mathematics offers absolute truth. The static and unalterable mode of presentation of concepts in the mathematics curriculum, rather than lack of knowledge, contributes to this misconception. Mathematical concepts, even the most elementary ones, have changed completely and repeatedly over time. Major contributions to the development of mathematics have been possible only because of significant revisions and expansions of the scope and contents of the objects of mathematics. Yet, we do not find this reflected in classroom teaching. While the room for integrating philosophy in mathematics education is very limited, an emphasis on the understanding of mathematical concepts is a necessary condition for a philosophical discourse about mathematics. The conceptual history of mathematics provides ample material for such focus, and leads to a better understanding of mathematics and our knowledge of mathematics. I will argue for the integration of the history of mathematics within the mathematics curriculum, as a way to teach students about the evolution and context-dependency of human knowledge. Such a view agrees with the contextual approach to rationality as proposed by Batens (2004). As a prime example, I will treat the development of the concept of a symbolic equation before the seventeenth century. In line with Lakatos (1976) and Kitcher (1984), my example is motivated by the epistemological relevance of the history of mathematics.

2. LIVING WITH INCONSISTENCIES

When asked for an example of an absolute truth, a student might likely answer "one plus one equals two." This is a grateful example to expand on. One plus one equals two is a current axiomatization of arithmetic, and is therefore true *with respect to that theory*. It is rather easy, however, to tailor the axiomatization to undermine the truth value of the given statement.

Adapting the Peano axioms leading to one being the successor of one, would yield the example false in the new theory.[3] Given that "one plus one equals two" is true in one theory and not in another refutes the example as an absolute truth. The student might object that changing the rules of arithmetic would lead to complete anarchy in society. The more intelligent student might notice that changing the Peano axioms in the given way would lead to an inconsistent theory, and that anything can be derived from inconsistencies. Let us look at these objections.

The point that changing the truth value of the given example makes no sense might be true, for now. There can be reasons, however, for changing the axioms of arithmetic. Van Bendegem (1994) did develop an inconsistent arithmetic by changing the Peano axioms so that there exists one number that is the successor of itself. His reason for doing so is to demonstrate the feasibility of a strict finite arithmetic. The fifth Peano axiom states that if equality applies for $x = y$ then x and y are the same number. This is the axiom that is tweaked by Van Bendegem so that starting from some number n, all its successors will be equal to n. If we take n to be one, then in this newly defined arithmetic, $1 + 1 = 1$. That would be a trivial arithmetic, however, which is not the intention of this enterprise. Rather than using one, the number n can be any number you like. Given a sufficiently large n, all operations of arithmetic behave the same way, as long as this number n is not reached during calculations. Now, a problem arrives when we reach n. The statement $n = n + 1$ is thus both true and false at the same time. This makes the new arithmetic inconsistent.

In classical logic you have the rule *ex falso quodlibet* (EFQ) which states that $p \wedge \neg p \rightarrow q$ or from an inconsistency you can derive anything. This would render the arithmetic trivial within classical logic (CL). Several paraconsistent logics now exist that do not have this problem, as well as inconsistency-adaptive logics developed at the Centre of Logic and Philosophy of Science (Batens 2001). Van Bendegem used the three-valued paraconsistent logic PL from Priest (1987), in which EFQ does not hold. With this underlying logic, he proved that if A is a valid statement in classical elementary number theory, then A is also valid in an elementary numbers theory based on a finite model. Gödel proved that every consistent formal theory that is rich enough to model arithmetic will contain true statements that cannot be proved within that theory. In other words, every consistent formal theory is incomplete. Giving up consistency, this new arithmetic, based on a finite model, has the advantage of being complete.

[3] The first axiomatization of arithmetic was given by Giuseppe Peano in a Latin publication of 1889, *Arithmetices principia, nova methodo exposita*. For an annotated English translation, see van Heijenoort 1967, 83–97.

There remains the objection of anarchy. What would happen if some people decided to change the rules of arithmetic? Would our accounting and wage calculation programs become unreliable when working with inconsistent arithmetic? In some sense, we already use this finite and inconsistent arithmetic in computer programs. An unsigned integer in a programming language such as C is represented by a 32- or 64-bit data structure, depending on the underlying hardware. Our inconsistent number n here becomes $2^{32} - 1$ or $2^{64} - 1$, while its successor is 0. Usually compilers warn for overflow situations such as these. When manipulating the binary structure with bit shift operations, the programmer has to reason within an inconsistent arithmetic and take care of the borderline situations himself. Apparently, many are more worried about giving up absolute certainty in mathematics than they are about their own life by relying on computers in daily situations. We do not have the slightest proof that the current commercial computers and compilers we use to create programs function the way we think they do. Such programs activate the anti-braking system in our car, guide traffic lights, and are used to calculate the structure of bridges and buildings. If they fail to work, human life may be at risk. There are attempts to prove the correctness of hardware design and computer programs, but these are not for practical or commercial use. In fact, we have proof of the contrary. Commercial computers have been known to be inconsistent in their arithmetic, as was shown with the famous Intel Pentium bug.[4] The fact, therefore, is that we live with inconsistencies every day of our life. Why is it so hard to accept this on a philosophical level?

3. ABSOLUTE CERTAINTY IN MATHEMATICS?

> "Gentleman, that $e^i + 1 = 0$ is surely true, but it is absolutely paradoxical; we cannot understand it, and we don't know what it means, but we have proved it, and therefore we know it must be the truth."

This well-known quote by Benjamin Peirce after proving Euler's identity in a lecture, reflects the predominant view of mathematicians before 1930, when mathematical truth equalled provability.[5] When Gödel proved that there are true statements in any consistent formal system that cannot be

[4] Given the calculation $x - \left(\frac{x}{y}\right) y = z$, the first Pentium chip produced the solution $z = 256$ for $x = 4195835$ and $y = 3145727$, instead of the correct $z = 0$. For more, see Coe, Tim, et al. 1995.

[5] Quoted in Kasner and Newman 1940.

proved within that system, truth became peremptorily decoupled of provability.

Peirce seems to imply something stronger, however, proving things in mathematics leads us to *the truth*. This goes beyond an epistemological viewpoint and is a metaphysical statement about the existence of mathematical objects and their truth independent of human knowledge. The great mathematician Hardy formulates it more strongly (Hardy 1929):

> "It seems to me that no philosophy can possibly be sympathetic to a mathematician which does not admit, in one manner or another, the immutable and unconditional validity of mathematical truth. Mathematical theorems are true or false; their truth or falsity is absolute and independent of our knowledge of them. In some sense, mathematical truth is part of objective reality."

Such statements are more than innocent metaphysical reflections open for discussion. They hide implicit values about the way mathematics develops, and have important consequences for the education and research of mathematics. An objective reality implies the fixed and timeless nature of mathematical concepts. The history of mathematics provides evidence of the contrary. Mathematical concepts—even the most elementary ones, like the concept of number—continuously change over time. The objects signified by the ancient Greek concept of *arithmos* differ from that of *number* by Renaissance mathematicians, which in turn differ from our current view. One could object that—not mathematics itself—our understanding of mathematical reality changes. Jacob Klein's landmark study (1934-6), however, focuses precisely on the *ontological* shift in the number concept. In Greek arithmetic *one* was not a number, but later it was. After that, the root of two was accepted as a number, and by the end of the sixteenth century, the root of -15 became a number.

4. LOOKING BEHIND THE BARRIER OF SYMBOLIC THINKING

Dealing with the development of symbolic algebra, we must define some terms more explicitly. Let us call algebra *an analytical problem-solving method for arithmetical problems in which an unknown quantity is represented by an abstract entity*. There are two crucial conditions in this definition: *analytical*, meaning that the problem is solved by considering some unknown magnitudes hypothetical and deductively deriving statements so that these unknowns can be expressed as a value, and an *abstract entity* that is used to represent the unknowns. This entity can be a symbol, a figure,

or even a color as, we shall see below. More strictly, symbolic algebra is *an analytical problem-solving method for arithmetical and geometrical problems consisting of systematic manipulation of a symbolic representation of the problem*. Symbolic algebra thus starts from a symbolic representation of a problem, meaning something more than a short-hand notation. There is no room here to expand on this important difference.[6] Instead, we will focus on one important misunderstanding: "as arithmetical problems are solved algebraically for over 3,000 years, an algebraic equation is a very old concept." This is not the case, as we shall argue. The symbolic equation is an invention of the sixteenth century.

We are all educated in the symbolic mode of thinking, which is so predominant that it becomes very difficult to grasp how non-symbolic algebra really works. In fact, in the history of mathematics there are many cases in which one completely ignored the difference. Let us take one example of Babylonian algebra. That Babylonians had an advanced knowledge of algebra is a fact that became known rather late—around 1930. Many thousands of clay cuneiform tablets were found that contained either tables with numbers or the solutions to numerical problems. One such tablet is YBC 6967 from Yale University, written in the Akkadian dialect around 1500 B.C. The most prominent scholar who studied and edited these mathematical tablets was Otto Neugebauer (1935-7, 1945). For the problem on YBC 6967, Neugebauer writes the following:[7]

> The problem treated here belongs to a well-known class of quadratic equations characterized by the terms igi and igi-bi (in Akkadian *igūm* and *igibūm* respectively) (..) We must here assume the product
>
> $$xy = 60 \quad (0.1)$$
>
> as the first condition to which the unknowns x and y are subject. The second condition is explicitly given as
>
> $$x - y = 7 \quad (0.2)$$
>
> From these two equations, it follows that x and y can be found from

[6] Mahony (1980) is one of the few to clarify the distinction. See also my forthcoming "Sixteenth century algebra as a shift in predominant models."

[7] Neugebauer and Sachs (1945, 129-30). The Babylonians used the sexadecimal number system, in which a unit is represented by Neugebauer as 1,0. I have changed this to decimal numbers and added the reconstructed text fragments for easier reading, which leaves the problem text otherwise intact.

$$x, y = \sqrt{\left(\frac{7}{2}\right)^2 + 60} \pm \frac{7}{2}$$

a formula which is followed exactly by the text, leading to $x = 12$ and $y = 5$.

Important here is that Neugebauer claims that equations are *explicitly given* and that the problem is "found from a formula which is followed exactly by the text." There are not so many people around who can go back to the cuneiform text and are able to check this claim. Fortunately, Neugebauer added an English translation which allows us to perform the task.

For the *explicitly given* equation, we read "The *igibūm* exceeds the *igūm* by 7." This indeed corresponds with the equation (0.3). For the formula, we read "As for you – halve 7, by which the *igibūm* exceeded the *igūm*, and the result is 3.5. Multiply together 3.5 with 3.5 and the result is 12.25. To the 12.25, that resulted you add 60, and the result is 72.25. What is the square root of 72.25?—8.5. Lay down 8.5, its equal, and then subtract 3.5 (the *takīlum*), from the one and add it to the other. One is 12, the other 5 (12 is the *igibūm*, 5 the *igūm*)." Again, the text seems to correspond with the formula. There are two minor details here: the *lay down* part sounds a little strange in this context, and Neugebauer adds "we have refrained from translating *takīlum*," because no sense could be given to it.

Recently, Jens Høyrup (2002) published a book that completely overthrows the standard interpretation of Babylonian mathematics and adds a new one. For Høyrup, Babylonian algebra works with geometric figures. This went by completely unnoticed because no figures appear on the tablets. Høyrup's study, however, is very convincing and its importance for the history of mathematics cannot be overestimated. In this problem, the unknowns *igibūm* and *igūm*, are represented by the sides of a rectangle (Høyrup 2002, 55-6). The term *product* used by Neugebauer should be read as *surface*, *square root* as *equal side* or the side of a square surface and adding means appending in length. According to Høyrup, the term *takīlum* should be read as *make-hold*, or making the sides of a rectangle hold each other. Only within a geometrical interpretation, does it make sense to lay down something. Using a rectangle with sides *igibūm* and *igūm*, everything fits together. The *igibūm* is 7 longer than the *igūm*. Cutting that part in half leads us to Figure 1.

Figure 1: an example of the geometric algebra from the Babylonians.

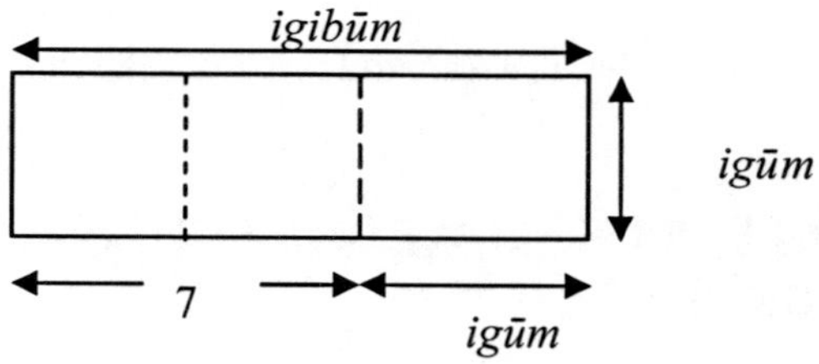

If we paste one of the halves below the rectangle at the length of the *igūm*, we get a figure with the same surface equal to 60.

Figure 2: Cut and paste method for solving quadratic problems.

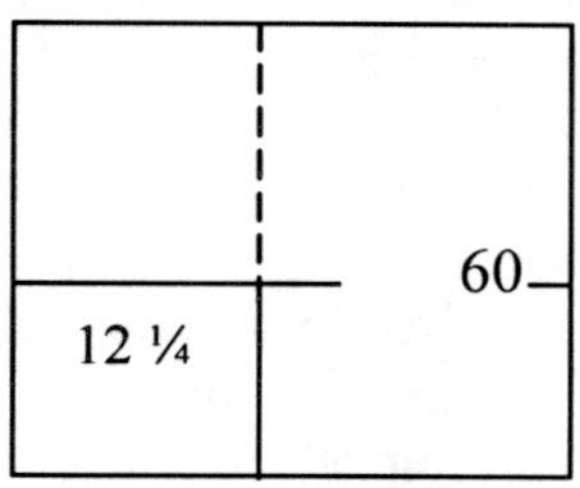

The part in the lower left corner must be a square, as its sides are both 3 ½. We can thus determine its surface as 12 ¼. The complete figure must also be a square with sides equal to *igūm* plus 3 ½. We know that the total surface is 72 ¼—the *equal side* of that square, therefore, is 8 ½. That leads us to a value of the *igūm* being 5. Pasting the cut-out half back to its original place gives a length of the *igibūm* of 12.

We are presented here with an interpretation completely different from that of Neugebauer. Høyrup accounts for anomalies in the standard interpretation and gives strong arguments for the reading of terms and actions in the geometrical sense. In this new interpretation, it makes no sense to speak about equations. Babylonian algebra does not solve equations, as the concept of an equation was absent. It fits in with our definition of algebra, however—the method is unquestionably analytical. It uses the unknowns *igūm* and *igibūm* and they are represented as abstract entities, namely the sides of a rectangle. We cannot blame Neugebauer for his symbolic reading of Babylonian algebra in 1945. Looking behind the barrier

of symbolic thinking proves to be a difficult task. His book was a major contribution to the early history of mathematics, but the history of mathematics has changed in the past decades and conceptual analyses such as Høyrup's have become the new methodological standard.

5. DIOPHANTUS: ALGEBRA OR THEORY OF NUMBERS?

The *Arithmetica* of Diophantus is often considered to be a primary source of European algebra.[8] This interpretation is questionable. The discovery of Diophantus in the fifteenth century had an important influence on the development of symbolic algebra. Its influence, however, is not as decisive as some want us to believe. The prime source for the myth that algebra was invented by Diophantus is Regiomontanus in his Padua lecture of 1464. Just having discovered the manuscript, Regiomontanus describes the *Arithmetica* enthusiastically as a book "in which the flower of the whole of arithmetic is hidden, namely the art of the thing and the *census*, which today is called algebra by an Arabic name. Here and there, the Latins have come in contact with this beautiful art[9]". Later, humanist mathematicians of the sixteenth century, such as Petrus Ramus, were more explicit with the idea that algebra originated with Diophantus and the Arabs learned the art from him.[10] Paradoxically, sixteenth-century humanists continued the program of reassessing mathematics from ancient sources, initiated by the Arabs, and in doing so precisely denied the contribution of the Arabs. Høyrup (1998) traces this evolution over several authors in Renaissance Europe. After Ramus, Bombelli, and Viète were also well acquainted with the *Arithmetica* and carefully avoided references to Arab influences. On the other hand, the Arab roots of algebra have mostly been acknowledged by the Italian abacus tradition from Fibonacci (1202, Boncompagni 1857) through the fifteenth century up to Cardano (1545, Witmer 1968) and the German *cossist* tradition, with Stifel (1544) as a most important author. It is probably thanks to them that we still use the name *algebra* today.

To assess the *Arithmetica,* it is important to draw a distinction between the context of the original text and its adaptations since its discovery by Regiomontanus. The treatment of problems from the *Arithmetica* by Bombelli (1572) and Simon Stevin (1585) are without doubt algebraic. Several editions of the *Arithmetica* have given an algebraic formulation to

8 E.g., Varadarajan 1991, Bashmakova 1997.

9 Regiomontanus 1972, 47, cited and translated by Høyrup 1998, 30.

10 Ramus gives a short history of mathematics in his *Scholae mathematicae* (1569).

problems, as has been done with Euclid's *Elements*. Such reformulation has been historically important for diophantine analysis, but was not necessarily a correct interpretation of the original work. Let us look at Problem 16 from the first book as an illustration. This is a rather simple problem looking for three numbers given their sum two by two:

Table 1: Two Interpretations of Problem 16 from Book I of the *Arithmetica* by Diophantus

Tannery (1893, 39)	**Ver Eecke (1926, 21, my translation)**
Invenire tres numeros tales ut bini simul additi faciant propositos numeros. Oportet propositorum trium dimidiam summam maiorem esse unoquoque horum.	To find three numbers which, taken two by two, form the proposed numbers. It is necessary, however, that half of the sum of the proposed numbers is larger than each one of these numbers.
Proponatur iam $X_1 + X_2 = 20$, $X_2 + X_3 = 30$, $X_3 + X_1 = 40$	Let us propose that the first number, increased with the second, forms 20 units; that the second, increased with the third, forms 30 units, and that the third, increased with the first, forms 40 units.
Ponatur $X_1 + X_2 + X_3 = x$	Let us pose that the sum of the three numbers is 1 *arithm*.
Quoniam $X_1 + X_2 = 20$, si a x aufero 20, habebo $X_3 = x - 20$	Consequently, because the first number with the second forms 20 units, if we take off 20 units from 1 *arithm*, we will have the third number: 1 *arithm* less 20 units.
Eadem ratione erit $X_1 = x - 30$, $X_2 = x - 40$	For the same reason, the first number will be 1 *arithm* less 30 units, and the second number will be 1 *arithm* less 40 units.
Linquitur summam trium aequari x, sed est haec summa $3x - 90$; ista aequentur x; fit $x = 45$.	It is necessary, still, that the sum of the three numbers becomes equal to 1 *arithm*. The sum of the three numbers, however, forms 3 *arithms* less 90 units. Let us equalize to 1 *arithm*, and the *arithm* becomes 45 units.
Ad positiones. Erit $X_1 = 15$, $X_2 = 5$, $X_3 = 25$. Probatio evidens est.	Let us return to what we posed: the first number will be 15 units, the second will be 5 units, the third will be 25 units, and the proof is clear.

Paul Tannery's respected critical edition of 1893 gives the original Greek text, reconstructed from several manuscripts, together with a Latin translation. As shown, the Latin translation presents the problem as one of three linear equations with three unknowns X_1, X_2, and X_3, and the use of an auxiliary x. The idea of linear equations with several unknowns, however, did not emerge before the mid-sixteenth century. Ver Eecke (1926) performed his French translation from the same Greek text as Tannery, but gives a more cautious interpretation. He does not use any symbols, and draws a distinction between number and *arithmos*. The unknowns X_1, X_2, and X_3 of Tannery are numbers in the French translation. Instead, the *arithmos* designates the unknown. After stating the problem, Diophantus reformulates the problem expressing the numbers in terms of a chosen unknown.

The interpretation of the *Arithmetica* as symbolic algebra is highly problematic. Even its designation as algebra cannot go without careful qualification. Nesselmann (1842) called it *syncopated algebra* as an intermediate stage between rhetoric and symbolic algebra. This would consist of short-hand notations that had not yet developed to full symbolism. The Greek text uses the letters Δ^{Υ} and K^{Υ}, which have been interpreted by many as the powers of an unknown, x^2 and x^3. Ver Eecke simply translates this as *square* and *cube* respectively. And this is without doubt closer to the original context than Tannery's Latin translation. Diophantus is primarily interested in the properties of numbers. A typical problem sounds like "Find two numbers with their sum and the difference of their squares given" (Book I, Problem 29; Tannery 1893, 65). The aim is to find numbers that satisfy the given property rather than solving the equations $x + y = 20, x^2 - y^2 = 80$. All problems of the *Arithmetica* are stated in the general way. A reading of the *Arithmetica* as a general theory of numbers is further emphasized by the character of diophantine problems having an infinity of numbers satisfying a given property. Diophantus's *Arithmetica* can be equally, or better, understood as a study on the properties of natural numbers than as early algebra. To read the text as an early form of symbolic algebra cannot be reconciled with the definition we have given above.

6. THE COLORFUL ALGEBRA OF THE HINDUS

Hindu tradition has passed down to us several important works on arithmetic and algebra, the importance of which to the development of algebra is still underestimated. The major handicap in drawing a line of influence of Indian sources on the development of Renaissance arithmetic and algebra is its

indirect character and lack of written evidence. We can trace some important paths of transmission for arithmetic and the Hindu-Arabic number system we currently use today. Some Arab texts that were translated into Latin clearly refer to Hindu sources.[11] Early arithmetic books are structurally very similar, for example, to Bhramagupta's *Brāhmasphuṭasiddhānta* (*BSS*) of 628 AD (Colebrook 1817). There is no known textual evidence, however, that shows a direct influence of Hindu algebra in the West. Comparing many problems treated in Hindu sources as well as in Renaissance algebra, we cannot avoid the particular similarity of both the formulation of the problems and most of the solution methods. Many linear problems solved algebraically in the abacus tradition have their counterpart in Hindu sources, while they are only rarely treated in Arab texts. We can discern an important influence of the oral tradition of recreational problems. Practical and recreational problems have functioned as vehicles for problem prototypes with typical solution patterns.[12] The solution method for typical problems are given as rules in Hindu texts. These rules are mostly formulated in Sanskrit verse, as stanzas or *sūtras*. Given the scarcity and cost of writing aids, memorizing aids in the form of verse was very important in mathematical texts before the age of printing. As an example, consider the following rule for solving linear problems given both in the *BSS* and the *Bīja-Ganita* (*BG*) of Bhāskarācārya of c. 1150:

Table 2: Solving Linear Problems *BSS* and *Bīja-Ganita* (*BG*) of Bhāskarācārya

Colebrook (1817, 227)	**Dvivedi (1902)**
Subtract the first colour [or letter] from the other side of the *equation*; and the rest of the colours [or letters] as well as the known quantities, from the first side: the other side being then divided by the [coefficient of the] first, a value of the first colour will be obtained. If there be several values of one colour, making in such case equations of them and dropping the denominator, the values of the rest of the colours are to be found from them.	Removing the other unknowns from [the side of] the first *unknown* and dividing the *coefficient* of the first *unknown*, the value of the first *unknown* [is obtained]. In case of more [values of the first unknown], two and two [of them] should be considered after reducing them to common denominators.

[11] *Dixit Algorizmi* c. 825. For a French translation, see Allard, 1992, 1-22.

[12] I have argued this more extensively in "How algebra spoiled Renaissance's practical and recreational problems" (forthcoming).

In the English rendition of the Sanskrit verses, Colebrook uses the term *equation* but he is not followed by Dvivedi. Instead, Dvivedi uses the terms *unknown* and *coefficient*, which in turn are not used by Colebrook. We can, therefore, cast some doubt about the use of these modern terms. Furthermore, Datta and Singh (1962, II, 9) claim that "in Hindu algebra there is no systematic use of any special term for the coefficient."

Prthūdakasvāmī (860), Srīpati (1039) and later Bhāskara (1150), solved linear problems by the use of several colors representing the unknowns. In other cases, flavors such as sweet (*madhura*) or flowers were also used for the same purpose. Solutions were mostly based on rules for prototypical cases, such as the rule of concurrence (*sankramana*) $\{x+y=a, x-y=b\}$, or the pulverizer (*Kuttaka*) $ax-by=c$. In several texts, starting with the *BSS*, we find reference to *samīcarana*, *samīcarā*, or *samīcriyā* often translated as *equation*. The rationale for this is that *sama* means *equal* and *cri* stands for *to do*. As with the terms *aequatio* and *aequationis* in early Latin works on algebra, we should be careful interpreting these terms in the modern way. They basically mean *the act of making even*—an essential operation in the algebraic solution of problems. They do not necessarily mean an equation in the sense of symbolic algebra. The basis of Hindu algebra is to reduce problems to the form of given precepts that provide a proven solution to the problem. The method is algebraic, as it uses abstract entities for the unknowns and is analytical in its approach. The Hindu methods for solving linear problems were transmitted to the West by prototypical problems, mostly of the recreational type, which served as vehicles for the corresponding problem-solving recipes. An example is the case $\{x+a=c(y-a), y+b=d(x-b)\}$, which we find in the *Ganitasārasangraha* of Mahāvīra and the *BG*, but also in several fifteenth-century arithmetics under the name *regula augmentationis*.

7. ARAB ALGEBRA

Arab algebra was introduced in Europe by the translations of the *Algebra* of Mohammed ibn Mūsa al-Kwārizmī by Guglielmo de Lunis, Gerhard von Cremona (1145), and Robert of Chester (1450; Hughes 1981). Most importantly however was the *Liber Abaci* of Fibonacci (1202). Fibonacci devoted the last part of his book to algebra and used mostly problems and solution methods from al-Kwārizmī and Al-Karkhī. Although Arab algebra developed to a high degree of sophistication during the next centuries, it was mostly the content of these early works that were known in Europe. Recent studies have provided us with a new picture on the continuous development

of algebra in the Italian abacus schools between Fibonacci and Luca Pacioli's *Summa de arithmetica geometria proportioni* (1494) (Franci and Rigatelli 1985). It took about four centuries before the transition to symbolic algebra was completed.

al-Kwārizmī gives solutions to algebraic problems by applying proven procedures in an algorithmic way. The validity of the solution is further demonstrated by geometrical diagrams. In contrast with Babylonian algebra, the method is not geometrical in nature—only the demonstration and interpretation is. As an example let us look at the way al-Kwārizmī solves Case 4 of the quadratic problem that can be represented by the well-known equation, $x^2+10x=39$ (Rosen 1831, italics are mine):

> For instance, one square, and ten roots of the same, amount to thirty-nine dirhems. That is to say, what must be the square which, when increased by ten of its own roots, amounts to thirty-nine? The solution is this: *you halve the number of the roots*, which in the present instance yields five. *This you multiply by itself*; the product is twenty-five. *Add this to thirty-nine*; the sum is sixty-four. *Now take the root of this*, which is eight, and *subtract from it half the number of the roots*, which is five; the remainder is three. *This is the root of the square which you sought for*; the square itself is nine.

If we write the case as $x^2+bx=c$, the solution fully depends on the application of the procedure that corresponds to $\sqrt{\left(\frac{b}{2}\right)^2+c}-\frac{b}{2}$. Solving problems in Arab algebra consists of formulating the problem in terms of the unknown and reducing the form to a known case. Methods for solving quadratic problems were given before in Babylonian and Hindu algebra. Again, we see no equations in Arab algebra. The explicit treatment of operations on polynomials is new, however. The basic operations of addition, subtraction, multiplication, and division, which were applied before to numbers, are now extended to an aggregation of algebraic terms. A further expansion of these operations would lead to the concept of a symbolic equation in the sixteenth century.

8. THE EMERGENCE OF THE CONCEPT OF AN EQUATION

So, what is it that constitutes the concept of an equation? I propose to adopt an operational definition of the term to reconstruct the historical emergence of the concept. We now consider an equation as a mathematical object on which certain operations are allowed. Let us, therefore, look at the precise

point in time in which an equation is named, consistently used, and operated upon as a mathematical object. As said before, the use of the term *aequatio* is not a sufficient condition for the existence of an equation. The observation that two polynomials are numerically identical does not in itself constitute an equation. An operation on an equation, however, would be. The first historical instance that I could find is in Cardano's *Practica arithmetice* (1539, f. 91^r).

This is probably the most important page in the development of symbolic algebra, as it combines two important conceptual innovations in a single problem solution—the use of a second unknown and the first operation on an equation.

Figure 3: The first operation on an equation in Cardano's *Practica arithmetice* of 1539.

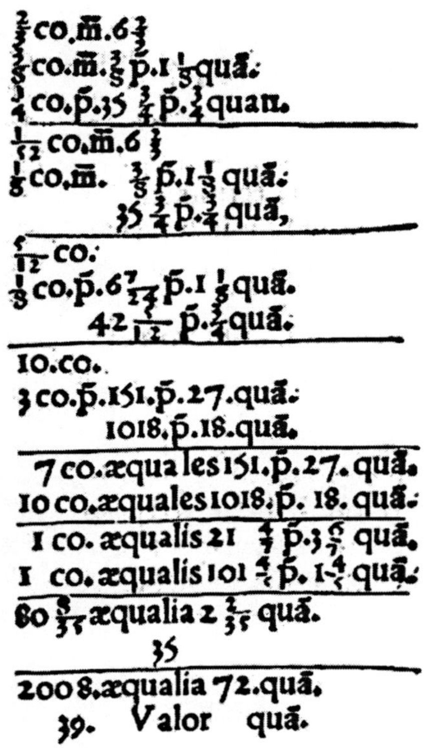

Cardano uses *co.* for the primary unknown and *quan.* for a secondary one. We can justly write this as x and y without misinterpreting the original context. In the example given in Figure 3, Cardano manipulates several polynomials, but at some point moves to equations.

We find 7 co. aequales 151 p. 27 quan. ($7x = 151 + 27y$) and 10 co. aequales 1018 p 18 quant ($10x = 1018 + 18y$). He divides these equations by 7 and 10 respectively, without explicitly saying so. By equating both, however, he arrives at $80\frac{8}{35} = 2\frac{2}{35}y$, which he explicitly multiplies by 35 to arrive at $72y = 2808$ or $y = 39$ (misprinted as $2008 = 72y$). From this moment on, algebra changed drastically. Cardano's book was widely read and several authors built further on this milestone. Stifel (1545) introduced the letters 1A, 1B, and 1C to differentiate multiple unknowns, which removed most of the ambiguities from earlier notations. It was Johannes Buteo (1559), however, who

established a method for solving simultaneous linear equations by systematically substituting, multiplying, and subtracting equations to eliminate unknowns. These developments between 1539 and 1559 constituted the concept of a symbolic equation. The equation became, not only a representation of an arithmetical equivalence, but also represented the combinatorial operations possible on the symbolic structure. This paved the road for Viète (1591) and Harriot to study the structure of symbolic equations.

9. CONCLUSION

We treated 3000 years of algebra in a few paragraphs with the risk of over simplification. One important conclusion emerges, however, —at some point in history there was a dramatic change in the way arithmetical problems were solved. By the second half of the sixteenth century, algebraic problem solving became the systematic manipulation of symbolic equations. We argue that the concept of an equation, as we understand it today, did not exist before that time. The development of sixteenth-century algebra is one of those occasions in which we see the birth of a new important concept in mathematics. Algebra did exist before, but functioned in a different way. Symbolic algebra, as it is currently taught in secondary education, is only one aspect of algebraic practice. While symbolic algebra may be the most efficient kind in problem solving, it is not always the most adequate one for teaching basic algebraic concepts to children. Luis Radford (1995, 1996, 1997) has demonstrated how procedures from the pre-symbolic abacus tradition can contribute to a better didactic understanding of the use of multiple unknowns. Joëlle Vlassis (2002) points out that conceptual difficulties with negative numbers originate from symbolic algebra and argues for a non-symbolic way to convey the concept. Our current conceptualization of algebra may confuse students in their first exposure to symbolic problem solving. An approach to improving such a situation, which has found some recognition during the past years, is to employ a plurality of methods. A new concept, method, or theorem, explained in multiple ways, is more likely to reach a broader range of students. Some students have difficulties with purely symbolic accounts of mathematics. Others are weak in spatial representations. Still others need numerical examples to be able to grasp abstract relations and functions. Teaching concepts by a plurality of methods levels out these difficulties. The history of mathematics provides a vast repository of alternative cases, representations, and methods.

An additional consequence of a plurality of methods and conceptualizations is situated on the philosophical level. If one thing should

be clear from our 3,000-year overview of algebra, it is that mathematics is subject to a historical process, that is based not only on the insights of some inventive individuals but also involves socio-cultural aspects of mathematical practice. The predominant view passed on by mathematics education hides implicit values on the superiority of modern ideas over past ones, and possibly of Western concepts over non-Western ones. Again, the history of mathematics shows that mathematics has always adapted to the needs of society. Mathematics was born in the Fertile Crescent, extending to the belt from North Africa to Asia, where wild seeds were large enough and mammals capable of employable domestication.[13] Modern algebra fertilized in the mercantile context of merchants and craftsman in Renaissance Italy. Several important figures in the development of symbolic algebra wrote also on bookkeeping, as well as on algebra often in one and the same volume.[14] If we accept that double-entry bookkeeping emerged in the fifteenth century as a result of the expanding commercial structures of sedentary merchant in Renaissance Italy, why not consider symbolic algebra within the same context? Ideas should be interpreted within the historical context in which they emerged and perhaps their superiority is dependent on the degree in which they adapted to the needs of society.

The idea of an objective reality of mathematical concepts, therefore, evades the veracity of conceptual dynamics and conceptual problems in mathematics. Conceptual dynamics challenges mathematical realism with some awkward questions. Take the simple concept of an algebraic equation as we have explored it. Of the different historical meanings of an equation, which one corresponds with the metaphysical object separate from human mathematical practice and understanding? If there have been different meanings for a given concept, such as an algebraical unknown, do they all correspond with the *object of an unknown* for a realist, or only our current conceptualization? If so, what is the ontological status of historical conceptualizations? What about inconsistent conceptualizations? Time and again, there have been serious crises in the conceptual foundations of mathematics.[15] There have been inconsistent theories, such as the early use of analysis and set theory, which have existed for several decades. It is precisely in times of crisis and conceptual difficulties that new ideas emerge

[13] For an eye-opening study on the relationship between these coincidental factors and the development of culture and thus mathematics, see the excellent work of Jared Diamond, 1996.

[14] Between 1494 and 1586: Luca Pacioli, Grammateus, Valentin Mennher, Elcius Mellema, Nicolas Petri and Simon Stevin.

[15] An important case study on crisis in mathematics is Carl Boyer, *The History of the Calculus and Its Conceptual Development*, 1959. As the title suggests, Boyer concentrates on the conceptual difficulties in developing the modern ideas of calculus.

and breakthroughs are made. According to Lakatos (1976, 140) such periods are "the most exciting from the historical point of view and should be the most important from the teaching point of view."

The aim of this paper is to show that the history of mathematics offers ample opportunities to illustrate the plurality of methods and the dynamics of concepts in mathematics. Integrating threads of conceptual development of mathematics in classroom teaching contributes to students' philosophical attentiveness. Such examples will alert students of the relativity of mathematical methods, truth, and knowledge and will put mathematics back in the perspective of time, culture, and context. The conceptual history of mathematics provides excellent opportunities to convey the basic epistemological and ontological questions of philosophy of mathematics in mathematics education.

REFERENCES

Allard, André (1992). *Muhammad Ibn Mūsā Al-Khwārizmī. Le Calcul Indien (Algorismus).* Paris: Blanchard.

Bashmakova, Isabella Grigoryevna (1997). *Diophantus and Diophantine Equations.* Dolciani Mathematical Expositions, 20, Washington DC: The Mathematical Association of America.

Batens, Diderik (2001). "A General Characterization of Adaptive Logics", *Logique et Analyse*, No. 173-174-175, (appeared 2003), 45-68.

Batens, Diderik (2004). *Menselijke Kennis. Pleidooi voor een bruikbare rationaliteit* 2nd edition, Antwerpen/Apeldoorn: Garant.

Bombelli, Rafaello (1572). *L'algebra: parte maggiore dell'arimetica: divisa in tre libri*, Bologna: Giovanni Rossi.

Boncompagni, Baldassarre (1857-62). *Scritti di Leonardo Pisano, matematico del secolo decimoterzo, Liber Abbaci di Leonardo Pisano.* 2 vols. Vol I., Rome: Tipografia delle Scienze Matematiche e Fisiche.

Boyer, Carl B. (1959). *A History of the Calculus and Its Conceptual Development.* New York: Dover, Second edition, 1989.

Buteo, Johannes (1559). *Logistica, quae et arithmetica vulgo dicitur, in libros quinque digesta...*, Lugduni: apud G. Rovillium.

Galinger, Ronald (Ed.) (1996). *Vita Mathematica. Historical Research and Integration with Teaching*, West Point, NY: The Mathematical Association of America.

Cardano, Girolamo (1539). *Practica arithmetice, & mensurandi singularis. In qua que preter alias cōtinentur, versa pagina demonstrabit*, Milan: Io. Antonins Castellioneus medidani imprimebat, impensis Bernardini calusci.

Coe, Tim, et al. (1995). "Computational aspects of the Pentium affair", *IEEE Computational Science and Engineering* (spring), 18-30.

Colebrook, Henry Thomas (1817). *Algebra with Arithmetic and Mensuration from the Sanscrit of Brahmegupta and Bháscara.* London: John Murray. Vaduz, Lichtenstein: Sandig Reprint Verlag, 2001.

Datta, Bibhutibhusan and Avadhesh Narayan Singh (1962). *History of Hindu Mathematics, a Source Book.* Bombay: Asia Publishing House (Reprinted in 2 vols., Delhi: Bharatiya Kala Prakashan, 2001).

Diamond, Jared (1996). *Guns, Germs, and Steel. The Fate of Human Societies.* New York: W. W. Norton.

Dvivedi, Sudhākara (1902). *The Brāhmasphuṭasiddhānta and Dhyānagrahapadeṣādhyāya*, Benares: Printed at the Medical Hall Press.

Flauvel, John and Jan van Maanen (Eds.) (2000). *History in Mathematics Education. The ICMI Study*, Dortrecht: Kluwer.

Franci, Raffaella and LauraToti Rigatelli (1985). "Towards a history of algebra from Leonardo of Pisa to Luca Pacioli", *Janus*, 72, nr. 1-3, 17-82.

Hardy, G. H. (1929). "Mathematical proof", *Mind* 38, 4, 1-25.

Høyrup, Jens (1998). "A new art in ancient clothes. Itineraries chosen between scholasticism and baroque in order to make algebra appear legitimate, and their impact on the substance of the discipline", *Physis*, Nuova Ser. 35, No.1, 11-50.

Høyrup, Jens (2002). *Lengths, widths, surfaces. A Portrait of Old Babylonian Algebra and Its Kin.* Heidelberg: Springer Verlag.

Hughes, Barnabas B. (1981). *Robert of Chester's Latin Translation of Al-Khwārizmī's Al-Jabr.* Stuttgart: Steiner.

Kasner, Edward and James Roy Newman (1940). *Mathematics and the imagination.* New York: Simon and Schuster.

Kitcher, Philip (1984). *The Nature of Mathematical Knowledge.* Oxford: Oxford University Press.

Klein, Jacob (1934-1936). *Die griechische Logistik und die Entstehung der Algebra.* Quellen und Studien zur Geschichte der Mathematik, Astronomie und Physik. Berlin. English translation: Greek mathematical thought and the origin of algebra, Massachusetts: MIT Press, 1968.

Lakatos, Imre (1976). *Proofs and Refutations.* Cambridge: Cambridge University Press.

Mahony, Michael S. (1980). "The Beginnings of Algebraic Thought in the Seventeenth Century", in S. Gaukroger (Ed.), *Descartes: Philosophy, Mathematics and Physics.* Barnes and Noble Books, Chap.5.

Priest, Graham (1987). *In Contradiction: A Study of the Transconsistent*, Dordrecht: Martinus Nijhoff.

Nesselmann, Georg Heinrich Ferdinand (1842). *Versuch einer kritischen Geschichte der Algebra. Vol 1. Die Algebra der Griechen*, Berlin, (Reprint Frankfurt: Minerva, 1969).

Neugebauer, Otto (1935-7). *Mathematische Keilschrifttexte*, 3 vols. Berlin: Springer.

Neugebauer, Otto and Abe Sachs (1945). *Mathematical Cuneiform texts.* New Haven: American Oriental Society, Series 29, Reprinted in 1986.

Radford, Luis (1995). "Before the other unknowns were invented: didactic inquiries on the methods and problems of medieval Italian algebra", *For the Learning of Mathematics*, 15, (3), 28-38.

Radford, Luis (1997). "L'invention d'une idée mathématique: la deuxième inconnue en algèbre", *Repères*, 28, 81-98.

Radford, Luis and G. Guérette, (1996). "Quadratic equations: Re-inventing the formula. A teaching sequence based on the historical development of algebra", *Proceedings of the Quadrennial Meeting of the International Study Group on the Relations Between History and Pedagogy of Mathematics and Deuxième Université d'été Européenne sur l'Histoire et l'Épstémologie dans l'éducation Mathématique*, Braga, Portugal: Universidade do Minho,Vol. II, 301-8.

Ramus, Petrus (1569). *Scholarum mathematicarum libri unus et triginta*, Basel: Eusebius Episcopius.

Regiomontanus, Johannes (1972). *Opera collectanea*, F. Schmeidler (Ed.), Osnabrück: Otto Zeller.

Rosen, Frederic (1831). *The algebra of Muhammad ben Musa*, London: The Oriental Translation Fund, (Reprinted, Hildesheim: Georg Olms, 1986).

Stevin, Simon (1585). *L'arithmetique de Simon Stevin de Bruges: contenant les computations des nombres Arithmetiques ou vulgaires: aussi l'Algebre, avec les equations de cinc quantitez. Ensemble les quatre premier livres d'Algebre de Diophante d'Alexandrie, maintenant premierement traduicts en franc□ois.* A Leyde: De l'imprimerie de Christophe Plantin.

Stifel, Michael (1544). *Arithmetica Integra cum præfatione P. Melanchthonis*, Nürnberg: Johann Petreius.

Tannery, Paul (1893). *Diophanti Alexandrini opera omnia cum graecis commentariis. Diophanti quae exstant omnia continens, ed. et latine interpretatus est Paulus Tannery.* Volume I, Bibliotheca scriptorum graecorum et romanorum. Stuttgart: Teubner.

Van Bendegem, Jean Paul (1994). "Strict finitism as a viable alternative in the foundations of mathematics", *Logique et Analyse*, Vol.37, No.145, (appeared 1996), 23-40.

van Heijenoort, Jean (1967). *From Frege to Gödel. A Source Book in Mathematical Logic*, 1879-1931, Cambridge, Ma.: Harvard University Press.

Varadarajan, V. S. (1991). *Algebra in Ancient and Modern Times*. Providence: The American Mathematical Society.

Viète, François (1591). *Francisci Vietae In artem analyticem isagoge. Seorsim excussa ab Opere restituae mathematicae analyseos, seu algebra nova.* Turonis: apud Iametium Mettayer typographum regium.

Ver Eecke, Paul (1926). *Diophante d'Alexandrie. Les six livres arithmétiques et le livre des nombres plygones.* Œuvres traduites pour la première fois du Grec en Français. Avec une introduction et des notes par Paul Ver Eecke. Ouvrage publié sous les auspices de la Fondation universitaire de Belgique. Brughes: Desclée, De Bouwer et Cie.

Vlassis, Joëlle (2002). *Algèbre par des situations problèmes au début du secondaire*, Louvain-la-neuve: De Boeck education.

Witmer, Richard T. (1968). *Ars Magna or the Rules of Algebra*. Cambridge, Mass.: M.I.T. Press, Reprinted by Dover Publications, New York, 1993.

INTERLUDE 5

It may seem like a gigantic leap from sixteenth century mathematics, algebra in particular, the time when Albrecht Heeffer's paper ended, to present-day mathematics, and more specifically, the theory of probability as taught in the classroom, the topic of Carmen Batanero and Carmen Díaz's paper. We believe this is not the case, which explains why we have chosen to let Heeffer be followed by Batanero and Díaz.

To take a rather trivial example, many problems in probability theory can be easily reformulated as algebraic problems. Is this not how we typically judge elementary probabilities? The six sides of a die are equivalent, so they must all have the same probability of coming up, but the sum of all probabilities is 1, so the probability of any one side coming up has to be 1/6. In algebraic terms we are simply asking for a number p, such that $6 \times p = 1$. Trivial indeed!

In the preceding paragraph, however, we used one word that is really begging for trouble: *equivalent.* As soon as one reflects on the meaning of the word, one uncovers one problem after another. This is shown quite clearly by Batanero and Díaz when they go through the known attempts at formulating a convincing definition of probability. As one might expect, none of them are completely right (that is, if we happen to know what *right* here means.) Should one then be amazed that children in a classroom setting have difficulties understanding and applying a probability concept, as they show? Of course not.

But we now have used another word that is really begging for trouble: *convincing.* Does this not seem truly odd? Here we are in the framework of mathematics, where proof is the central idea and the core aim to strive for, and all of a sudden we are talking about *convincing someone.* Does not such a term belong within the humanities, where arguments and similar devices are the best tools one can use? In other words, proof belongs to the exact sciences, argument to the human sciences. Tempting though it may perhaps be to nod approvingly, one must resist if science wars are to be avoided. We are left with an important problem to deal with: what concept or theory can unite proofs and arguments, show their relatedness and their continuity? Perhaps not surprisingly the answer is: *semiotics.* For is not mathematics on a par with music, logic, and playing games, if we look at all of them as very particular sign systems, sometimes partly ritualized and partly formalized, and often used for creating, if not maintaining, social contrasts?

THE MEANING AND UNDERSTANDING OF MATHEMATICS

The Case of Probability

Carmen Batanero and Carmen Díaz
University of Granada, Spain

Abstract: We summarize a model with which to analyze the meaning of mathematical concepts, distinguishing five interrelated components. We also distinguish between the personal and the institutional meaning to differentiate between the meaning that has been proposed for a given concept in a specific institution, and the meaning given to the concept by a particular person in the institution. We use these ideas to analyze the historical emergence of probability and its different current meanings (intuitive, classical, frequentist, propensity, logical, subjective and axiomatic). We furthermore describe mathematical activity as a chain of semiotic functions and introduce the idea of semiotic conflict that can be used to give an alternative explanation to some widespread probabilistic misconceptions.

Key words: Probability, history of probability, proof, semiotics, misconceptions

1. INTRODUCTION

The teaching of probability has been included for many years in the mathematics curriculum for secondary school. There is, however, a recent emphasis on the experimental approach and on providing students with stochastics experience (e.g., M.E.C. 1992; N.C.T.M. 2000; Parzysz 2003). As argued in Batanero, Henry, and Parzysz (2005) these changes force us to reflect on the nature of chance and probability, since the analysis of obstacles that have historically emerged in the formation of concepts can help educators understand students' difficulties in learning mathematics.

Moreover, a well-grounded mathematics education researcher or teacher requires a wide view of understanding, so that understanding probability, for

example, is not simply reduced to the student's ability to define the word. Vergnaud (1982; 1990) suggested that psychological and educational researchers should consider a concept to include not just the set of invariant properties that make the concept meaningful, but also the situations (problems, task, phenomenology) and representations associated with the concept. Godino and Batanero (1999) take from Vygotski (1934) the suggestion that the meanings of words are the main units to analyze psychological activity, since words reflect the union of thought and language, and include the properties of the concept to which they refer. As such, one main goal in mathematics education research is finding out what meanings students assign to mathematical concepts, symbols, and representations; and explaining how these meanings are constructed during problem solving activities and how they evolve, as a consequence of instruction, and progressively adapt to the meanings we are trying to help students understand. Of particular relevance for the teaching of probability are the informal ideas which children and adolescents assign to chance and probability before instruction, which can affect their subsequent learning. For example, Truran (1995) found substantial evidence that young children do not see random generators such as dice or marbles in urns as having constant properties, but believe that a random generator has a mind of its own or may be controlled by outside forces.

1.1 Components of the Meaning and Understanding of Mathematics

In trying to develop a systematic research program for mathematics education at the University of Granada, Spain, we have developed a theoretical model to carry out these analysis (Godino and Batanero 1994; 1998; Godino 2002; Godino, Batanero and Roa 2005; Godino, Contreras, and Font in press), which has been successfully applied in different works of research in statistics education, in particular in some Ph.D. theses carried out at different universities in Spain.

In this paper we will use the concept of probability as an example, although the theory we are describing is also useful for other types of mathematical objects, such as theorems (e.g., the central limit theorem) or even a complete part of mathematics or statistics (e.g., variance analysis). In this model we distinguish five interrelated components in the meaning of the concept, as described below:

- *The field of problems from which the concept has emerged.* One such problem was posed to Galileo by the Great Duke of Tuscany (about 1620): although 9 and 12 can be made up as the sum of the eyes of two dice, in as many different ways as 10 and 11, and

therefore should be expected to have the same frequency, the observation of long series of trials makes players prefer 10 and 11 to 9 and 12. In spite of its simplicity both Leibnitz and D'Alembert were unable to solve this problem (Székely 1986). Many other problems related to chance games were used to develop the first ideas of expectation and probability.

- *The representations of the concept.* To solve problems we need ostensive representations, since concepts are abstract entities. For example, in his letter to Fermat dated July 29, 1654, Pascal used the words *value*, *chance*, *combinations*, as well as numbers, symbols (letters), fractions, and a representation of the arithmetic triangle to solve a problem proposed by the Chevalier de Meré. In another letter to Fermat dated August 24, 1654, he used a tabular arrangement to enumerate the different possibilities in an interrupted game (Pascal 1963/1654). Modern representations of probability include density curves, algebraic expressions, distribution tables, or dynamical graphs produced by computers.
- *The procedures and algorithms to deal with the problem and data, to solve related problems, or to compute values.* Primitive probability problems were solved by simple enumeration or using other combinatorial tools. Today we have a wide variety of mathematical tools to help us solve probability problems, including combinatorics, analysis, algebra, and geometry. Distribution tables, calculators, and computers have also reduced the gap between the understanding of a problem and the technical competence required to solve it (Biehler 1997).
- *The definitions of the concept.* These will include its properties and relationships to other concepts, such as the different definitions of probability, the idea that a probability is always positive, the sum and product rule, the relationships between probability, expectation, frequency, and odds, and limit theorems.
- *The arguments and proofs we use to convince others of the validity of our solutions to the problems or the truth of the properties related to the concepts.* Galileo gave a complete combinatorial proof of the solution of the two dice problem and showed by enumeration that there are 25 different possibilities for both 9 and 12 and 27 for both 10 and 11. This same proof would be understandable by secondary school students; although the teacher might first allow the students to get some experience with randomness by organizing a classroom experiment in which students would throw the dice, record the results, and compare the relative frequencies of the different sums after a long series of trials. Computers and Internet applets possibly

> could improve the simulation and increase the students' possibilities of exploring the experiments. This experimental confirmation is, however, very different from mathematical proof, although it can play a main role in the probability classroom, especially when we are dealing with complex ideas, such as sampling distributions.

Our previous discussion suggests the multifaceted nature of even apparently simple concepts, such as probability and the need to take into account the different *elements of meaning* in organizing instruction. It is also important to notice that different levels of abstraction and difficulty can be considered in each of the five components defined above, and that the *meaning of probability* is thus very different at different institutions. In primary school or for the ordinary citizen, an intuitive idea of probability and the ability to compute simple probabilities by using the Laplace rule would be sufficient, using a simple notation and avoiding algebraic formulae. A probability literate citizen (Gal, in press) would also need to understand the use of probability in decision making situations (stock market, medical diagnosis), sampling, and voting, etc. In scientific or professional work, or at university level, however, a more complex meaning of probability, including knowledge of main probability distribution, limit theorems, and even stochastic processes would be needed.

We therefore distinguish between *personal* and *institutional* meanings to take into account these varieties of meanings for the same concept at different institutions and also to differentiate between the meaning that has been proposed or fixed for a given concept in a specific teaching institution, and the meaning given to the concept by a particular student in the institution.

2. MEANINGS OF PROBABILITY

These ideas are particularly relevant in analyzing the historical emergence of probability and its different meanings (laplacian, frequentist, subjective, axiomatic, etc.). The concept of probability has received different interpretations according to the metaphysical component of people's relationships with reality (Hacking 1975), and the progressive development of probability has been linked to a large number of paradoxes that show the disparity between intuition and conceptual development in this field (Székely 1986, Borovcnik and Peard 1996). Below we summarize these different meanings, using some ideas from Batanero, Henry, and Parzysz (2005).

2.1 Intuitive Meaning

No one knows when humans first started to play games of chance. Nor is it clear why probability theory only started recently as compared to other branches of mathematics. Hacking (1975) analyzes and discards different reasons that have been proposed to explain this delay, including obsession with determinism, lack of collections of empirical frequencies, religious beliefs, scarcity of easily understood empirical examples, or economic incentives; and he argues that even if none of these reasons are valid, the idea of probability did not appear until around 1600.

Intuitive ideas related to chance and probability appeared very early, as they appear in young children or non-educated people who use qualitative expressions (probable, unlikely, feasible) to express their degrees of belief in the occurrence of random events. These ideas were of course too imprecise, and people needed the fundamental idea of assigning numbers to uncertain events to be able to compare their likelihood, thus applying mathematics to the wide world of uncertainty. Bellhouse (2000) analyzed a 13th-century manuscript, "De Vetula", attributed to Richard de Fournival (1201-1260), which is possibly the oldest known text establishing the link between observed frequencies and the enumeration of possible configurations in a game of chance. By counting the 216 possible permutations of all the different values of three dice, the author computes the possibilities of the 16 different sums.

Hacking (1975) indicates that probability has had a dual character since its emergence: a statistical side is concerned with stochastic rules of random processes, while the epistemic side views probability as a degree of belief. This duality was present in many of the authors who contributed to the progress of probability theory. For example, while Pascal's solution to games of chance reflects an objective theoretical probability, his theoretical argument for the existence of God is a question of personal belief.

2.2 Classical Meaning

The first probability problems were linked to games of chance, and it is therefore not surprising that the pioneer interpretations of probability were expressed in terms of winning expectations. Cardano (1961/1663) advised players in his "Liber de Ludo Aleae" to consider the number of total possibilities and the number of ways the favorable results can occur, and compare the two numbers in order to make a fair bet. Pascal (1963a/1654) solved the problem of estimating the fair amount to be given to each player in an interrupted game by proportionally dividing the stakes among each player's chances. In his "Traité du triangle arithmétique" (1963b/1654) he

developed combinatorial rules that he applied to solve some probability problems arising in his correspondence with Fermat. In "De Ratiociniis in Aleae Ludo" Huygens (1998/1657) showed that if p is the probability of someone winning a sum a, and q that of winning a sum b, then one may expect to win the sum $pa + qb$.

The first definition of probability was given by Abraham de Moivre in "The Doctrine of Chances": *Wherefore, if we constitute a Fraction whereof the Numerator is the number of Chances whereby an Event might happen, and the Denominator the number of all the chances whereby it may either happen or fail, that Fraction will be a proper definition of the Probability of happening* (de Moivre, 1967/1718, 1).

This definition was not without problems. Laplace suggested that the theory of chance consists of reducing all the events of the same kind to a certain number of equally possible cases, that is to say, to such as we may be equally undecided about in regard to their existence, and gave this definition: *probability is thus simply a fraction whose numerator is the number of favourable cases and whose denominator is the number of all cases possible* (Laplace, 1985/1814, 28).

This Laplacian definition of probability was based on a subjective interpretation, associated with the need to judge the equipossibility of different outcomes. Although equiprobability is clear when throwing a die or playing a chance game, it is not the same in complex human or natural situations. As noted by Bernoulli in "Ars Conjectandi" published in 1713, equiprobability can be found in very rare cases and does not only happen except in games of chance.

2.3 Frequentist Meaning

Bernoulli suggested a possible way to assign probabilities to real events, in applications different from games of chance, through a frequentist estimate (Bernoulli 1987/1713). He also justified a frequentist estimation of probability in giving a first proof of the Law of Large Numbers: if an event occurs a particular set of times *(k)* in n identical and independent trials, then if the number of trials is arbitrarily large, k/n should be arbitrarily close to the "objective" probability of that event.

The convergence of frequencies for an event, after a large number of identical trials of random experiments, had been observed over several centuries. Bernoulli's proof that the stabilized value approaches the classical probability was interpreted as a confirmation that probability was an objective feature of random events. Given that stabilized frequencies are observable, they can be considered as approximate physical measures of this probability.

In the frequentist approach, moreover, probability is defined as the hypothetical number towards which the relative frequency tends when stabilizing (Von Mises 1952/1928). In this conception, the existence of the number for which the observed frequency is an approximate value is assumed. According to authors such as Gnedenko and Kolmogorov, mathematical probability would be a useless concept if it did not find this precise expression in the relative frequency of events resulting from long sequences of random trials, carried out under identical conditions.

From a practical viewpoint, however, the frequentist approach does not provide the exact value of the probability of an event, and we cannot find an estimate of the probability when it is impossible to repeat an experience a very large number of times. It is also difficult to decide how many trials are needed to get a good estimation for the probability of an event. And of course we cannot give a frequentist interpretation of the probability of an event which occurs only one time under the same conditions, such as is often found in econometrics (Batanero, Henry, and Parzysz, 2005).

2.4 Propensity Meaning

In trying to solve some of the problems in the frequentist meaning of probability, as well as to make sense of single-case probabilities, Popper (1957, 1959) considered probability as a physical *propensity*, disposition, or tendency to produce an outcome of a certain kind, or the long run relative frequency of such an outcome. This idea was also suggested by Peirce (1932/1910) who advanced a concept of probabilities according to which a die, for example, possesses *would-bes* for its various possible outcomes, and these would-bes are intentional, dispositional, directly related to the long run, and indirectly related to singular events.

While the classical and frequentist meanings reduce the notion of probability to other, already known concepts (ratio, frequency, etc.), Popper introduces the idea of *propensity* as a measure of the "probabilistic causal tendency" of a random system to behave in a certain way. This idea was discussed by different authors, who distinguished *long run* and *single case* propensity (Gillies 2000). In the long run theories, propensities are tendencies to produce relative frequencies with particular values, but the propensities are not the probability values themselves (Popper, 1957, 1959, Hacking 1965); for example, a fair die has an *extremely strong* tendency (propensity) to produce a 5 with long run relative frequency of 1/6. The probability value 1/6 is small, so it does *not* measure this strong tendency. In single-case theory (e.g., Mellor 1971) the propensities are identical to the probability value and are considered as probabilistic causal tendencies to

produce a particular result on a specific occasion. In this theory we consider the die to have a *weak* propensity (1/6) to produce a 5.

Again this interpretation was controversial. In the long run interpretation, propensity is not expressed in terms of other empirically verifiable quantities, and we then have no method of empirically finding the value of the propensity. Moreover, the assumption that an experimental arrangement has a tendency to produce a given limiting relative frequency of a particular outcome presupposes a uniformity which is hard to test, either a priori or empirically (Cabriá 1992).

As regards single case interpretation, it is difficult to assign an objective probability for single events. The probability will vary according different conditions, and then the probability is attached to the conditions and not just to the event itself. Since the same event can be included in different reference sets, we might introduce subjective elements when selecting the reference class in which to include the event (Gillies 2000). It is also unclear whether single case propensity theories obey the probability calculus or not.

2.5 Logical Meaning

Keynes (1921) and Carnap (1950), among other authors, developed the logical theories, which retain the classical idea that probabilities can be determined a priori by an examination of the space of possibilities, although the possibilities may be assigned unequal weights. This approach defines probability as a *degree of implication* that measures the support provided by some evidence E to a given hypothesis H. Keynes accepts the conventional calculus of probability since it applies to his probability relations. He symbolizes certainty and impossibility by 1 and 0, and all other degrees of the probability relation lie between these limits. Since the deductive relations of implication and incompatibility can be considered as extreme cases (with degree of implication values 1 and 0 respectively), deductive logic is amplified in this approach.

Carnap (1959) started by constructing a formal language and defined probability as a rational credibility function, using the technical term *degree of confirmation.* The degree of confirmation of one hypothesis H, given some evidence E, depends entirely on the logical and semantic properties of and relations between H and E, and therefore it is only defined for the particular formal language in which these relations are made explicit. This degree of confirmation is just the conditional probability of H given E and allows inductive learning from experience. Carnap reduces the problem of defining the degree of confirmation for the particular formal language in which he is operating to the problem of selecting a probability for elementary state descriptions in that language.

One problem in this approach is that there are many possible confirmation functions, depending on the possible choices of initial measures and on the language in which the hypothesis is stated and in which the confirmation function is defined. A change of language might then make invalid any particular confirmation of a theory. A further problem is in selecting the adequate evidence E in an objective way, since the amount of evidence might vary from one person to another.

2.6 Subjective Meaning

Bayes's formula permitted the finding of the probabilities of various causes when one of their consequences is observed. The probability of such a cause would thus be prone to revision as a function of new information and would lose its objective character postulated by the above conceptions. Keynes, Ramsey, and de Finetti described probabilities as degrees of belief, based on personal judgment and information about experiences related to a given outcome. They suggested that the possibility of an event is always related to a certain system of knowledge and is thus not necessarily the same for all people.

The difficulty with the subjectivist viewpoint is that it seems impossible to derive mathematical expressions for probabilities from personal beliefs. Both Ramsey (1926) and de Finetti (1937), however, suggested a way of deriving a consistent theory of choice under uncertainty that could isolate beliefs from preferences while still maintaining subjective probabilities. The basic idea behind the Ramsey-de Finetti derivation is that by observing the bets people make, one can assume that this reflects their personal beliefs about the outcomes, and then subjective probabilities can be inferred.

From this subjectivist viewpoint, the repetition of the same situation is no longer necessary to make sense of probability. The fact that repeated trials are no longer needed is used to expand the field of applications of probability theory. Today, the Bayesian school assigns probabilities to uncertain events, even non-random phenomena, although controversy remains about the scientific status of results which depend on judgments that vary with the observer.

2.7 Mathematical Meaning and Axiomatization

Throughout the 20th century, different mathematicians contributed to the development of the mathematical theory of probability. Borel's view of probability as a special type of measure was used by Kolmogorov, who applied sets and measure theories to derive a satisfactory axiomatic system,

which was generally accepted by different schools regardless of their philosophical interpretation of the nature of probability.

Probability is, therefore, a mathematical object, and probabilistic models are used to describe and interpret random reality. Probability theory proved its efficiency in many different fields, but the particular models used are still subjected to heuristic and theoretical hypotheses, which need to be evaluated empirically. These models also allow assigning probabilities to new events that made no sense in previous interpretations. One such example is the Cantor distribution, a probability distribution with a Cantor function as a cumulative distribution. This distribution is a mixture of continuous and discrete; it has neither probability density function nor point-probability masses. It serves to solve problems such as the following: Let the random variable X take a value in the interval [0, 1/3] if we get heads in flipping a coin and in the interval [2/3, 1] if we get tails. Let X then take a value in the lowest third of the aforementioned interval if we get heads in the second flipping of the coin, and in the highest third if we get tails. If we imagine this process continuing to infinity, then the probability distribution of X is the Cantor distribution, with an expected value of 1/2. Of course interest in this distribution is mainly theoretical.

2.8 Summary

In this brief description, we have analyzed different historical views of probability that still persist and are used in the teaching and practice of probability. We can consider these views as different meanings for probability, according to our theoretical framework, since there are differences not only in the definition of probability, but also in the related problems, tools, representations, properties, and concepts that have emerged to solve various problems (see a summary in Table 1).

When teaching probability (or any other mathematical concept), these five different types of knowledge should be considered and interrelated, as research on the understanding of mathematics has shown that students find difficulties in each of these components. In the same way, the different meanings of probability should be progressively taken into account, starting with the students' intuitive ideas of chance and probability, since, understanding is a continuous constructive process in which students progressively acquire and relate the different elements of the meaning of the concept.

Table 1. Elements Characterizing Different Historical Meanings of Probability

Meaning of Probability	Problem Field	Tools	Representations	Properties	Some Related Concepts
Intuitive	• Games of chance • Divination	• Random generators	• Colloquial expressions	• Unpredictable • Opinion, belief	• Luck • Fate
Classical	• Games of chance	• Combinatorics • Proportions • A priori analysis of the structure	• Pascal Triangle • Enumerations • Combinatorial formulas	• Quotient of favorable cases divided by number of all cases possible • Equiprobable simple events	• Expectation • Fairness
Frequentist	• Repeatible experiments	• A posteriori recording of statistical data • Fitting mathematical curves • Mathematical analysis	• Statistical tables/graphs • Simulations • Density curves • Random number tables • Distribution tables	• Limit of relative frequencies in long runs • Objective character, based on empirical facts	• Relative frequency • Universe • Random variables • Probability Distribution
Propensity	• Physical situations, including single cases	• A priori analysis of the experimental set-up		• Physical disposition or tendency • Objective character • Applicable to single cases • Related to the experimental conditions	• Propensity • Probabilisitic causal tendency • Virtual frequency
Logical	• Widening inferences	• A priori analysis of the space of possibilities • Propositional logic • Inductive logic	• Formal language • Notation of conditional probability •	• Objective degree of belief • Relationship between two statements • Possibilities can be given unequal weights • Generalizes implication • Revisable with experience	• Evidence • Hypothesis • Degree of implication
Subjective	• Uncertain events, even non-repeatable	• Bayes's theorem • Conditional probability decision theory •	• Notation of conditional probability • Graphs of prior and posterior distributions	• Subjective character • Revisable with experience	• Likelihood • Exchangeability • Utility
Axiomatic	• Random experiments	• Set theory and algebra • Algebra • Analysis	• Boolean diagrams and set notations	• Measurable function	• Sample space • Borel sets

3. SEMIOTIC FUNCTIONS AND MATHEMATICAL REASONING

While our analysis of the elements of meaning is used to focus the attention on the different components of mathematical teaching and learning, it is insufficient to describe mathematical reasoning and mathematical activity (at a micro level of analysis). From a didactical point of view it is useful to consider the notion of semiotic function: *"There is a semiotic function when an expression and a content are put in correspondence"* (Eco 1979, 83). The original in this correspondence is the significant (plane of expression), the image is the *meaning* (plane of content), that is, what is represented, what is referred to by the speaker.

An *elementary meaning* is produced when a person interprets a semiotic function with a semiotic act (interpretation or understanding), in which he/she relates an expression to a specific content. An elementary meaning is the content that the author of an expression refers to, or the content that the reader/listener interprets. Some examples are given below:

- The expression "product rule" or the symbolic representation $P(M \cap I) = P(M) \times P(I)$ describes a procedure carried out by a student to compute a compound probability in the case that M and I are independent.
- The statement of a problem can represent the real situation; the simulation of random phenomena can represent the phenomena themselves; for example, we can simulate the number of girls in a sample of 10 newborn babies with coins.
- We can say "Pascal and Fermat's solution of de Meré's paradox" to refer to an argument.
- In expressions such as, "Let μ be the mathematical expectation of a random variable", the notations μ, the expressions *mathematical expectation* and *random variable* refer to particular abstract concepts.

Semiotic functions are always involved in creating mathematical concepts, establishing and validating mathematical propositions, and, as a rule, in problem solving processes; and they can produce *elementary* or *systemic meanings*. For example, when we say "computing probabilities in the normal distribution" in a general sense we refer to the whole set of procedural elements to compute these probabilities, including use of tables, standardization, computer programs, simulation, etc. When we speak of studying the *normal distribution*, we refer to the whole system of practices

associated with this distribution, whose structural elements are the five types described in section 1. These are examples of systemic meanings.

In each semiotic function, the correspondence between the expression and the content is fixed by explicit or implicit codes, rules, habits, or agreements. Sometimes the teacher and the student attribute different meanings to the same expression and there is a *semiotic conflict* (Godino, 2002). These conflicts appear in the linguistic interaction between people or institutions, and they frequently explain the difficulties and limitations of teaching and learning mathematics.

3.1 Reasoning as a Chain of Semiotic Functions

When solving a mathematical problem or carrying out any mathematical activity or reasoning, one or more semiotic functions appear among the five components described in section 1. It is then useful to consider mathematical reasoning as a sequence of semiotic functions (or as a chain of related pieces of knowledge) that a person establishes in the problem-solving process. The notion of *semiotic function* is used to indicate the essentially relational nature of mathematical activity and of teaching/learning processes. Let us consider, as an example, the incorrect solution given by a student to the following problem (taken from Díaz and de la Fuente 2006).

Problem
We threw two dice and the product of the two numbers was 12. What is the probability that neither of the two numbers was 6?

The solution given by a student is the following:

$$3 \cdot 4 = 12 \qquad P(3_1 \cap 4_2) = \frac{1}{6} \cdot \frac{1}{6} = \frac{1}{36}.$$
$$4 \cdot 3 = 12 \qquad P(4_1 \cap 3_2) = \frac{1}{6} \cdot \frac{1}{6} = \frac{1}{36}. \qquad \Big\} \; \frac{2}{36}$$

We have broken down this solution into analysis units in order to show the semiotic activity carried out by the student and to discover his or her semiotic conflicts (see Table 2).

Table 2. Semiotic Analysis of the Student's Solution

Analysis Unit	Semiotic Functions Established by the Student
[U1] *3.4=12; 4.3=12*	The expressions 3, 4 (symbols) refer to the numbers 3 and 4 (concepts) and these refer to the possible outcomes for each die (phenomenological object). The expression 12 (symbol) refers to the number 12 (concept) and this refers to the fixed value of the product (operation). The student identifies the favorable cases in the event (applies a concept to a practical situation). The student is able to distinguish that order (concept) is relevant.
[U2] $P(3_1 \cap 4_2)=\frac{1}{6}x\frac{1}{6}=\frac{1}{36}$	The expression $P(3_1 \cap 4_2)$ refers to compound probability and at the same time to the intersection of two events (concepts). [U2] refers to the product rule (proposition) in the case of independent events (property). The expression $\frac{1}{6}x\frac{1}{6}=\frac{1}{36}$ refers to the product of fractions (rule) and its result (procedure). The student is able to perceive the independence of the two dice. The student distinguishes the order of outcomes.
[U3] $P(4_1 \cap 3_2)=\frac{1}{6}x\frac{1}{6}=\frac{1}{36}$	Similar to Unit 2
[U4] $> \frac{2}{36}$	The arrows (graphical representation) refer to the sum of the two probabilities. The symbol $\frac{2}{36}$ refers to the value of this sum. The student is correctly identifying and applying the union axiom to compute the probability of the union of two incompatible sets. The student does not restrict the sample space.

We can see from this example the complexity of solving even elementary probability problems. This student was able to identify the problem data, apply the sum and product rules, assess independence and the relevance of order, use complex notation, and identify the favorable cases. He has not, however, restricted the sample space in computing the conditional probability (none of the numbers is 6, given that the product is 12) to the relevant cases (only 4 possibilities). A semiotic conflict appears when the student computs the compound probability "that the product is 12 and none of the numbers is six," instead.

4. PROBABILITY MISCONCEPTIONS AND SEMIOTIC CONFLICTS

Probability misconceptions have been widely documented, although recently some researchers are wondering if they can be explained only in terms of psychological mechanisms (e.g., Callaert, in press). Below we analyze some of these misconceptions and suggest that semiotic conflicts are alternative explanations for some of them.

4.1 Equiprobability Bias

Pratt (2000) reports interviews with two children who declared that the totals 2 and 3 for the sum of two dice were equally easy to obtain. He explained this response in terms of Lecoutre's (1992) equiprobability bias, by which children consider all the events in a random experiment to be equiprobable. Callaert (2003) analyzed these results and remarked that the final outcome for which the probability is being determined is not at all related, in a simple and direct way, to any underlying experiment that can easily be conceptualised by the children. In the experiment there are exactly 36 different ordered pairs (x,y) of numbers and 21 different unordered pairs of numbers.

When the children in Pratt's (2000) experiment are interviewed about their response, one of them (Anne) replies: "Well, 1 and 2 and 2 and 1 are the same ...; they come to the same number". As suggested by Callaert, children were paying too much attention to the mathematical operation of adding numbers. Since this operation is commutative, it invites children not to pay attention to order when computing probabilities. Here a semiotic conflict appears when children assign a non–existent property (commutativity) to the sample space, assuming the outcomes (1, 2) and (2, 1) are identical when they are not. The students and the teachers are, moreover, assuming different sample spaces for this experiment (see Table 3, a and b).

Table 3. Different Sample Spaces Assumed by the Teacher and the Student

a. Sample Space Assumed by the Teacher

1,1	1,2	1,3	1,4	1,5	1,6
2,1	2,2	2,3	2,4	2,5	2,6
3,1	3,2	3,3	3,4	3,5	3,6
4,1	4,2	4,3	4,4	4,5	4,6
5,1	5,2	5,3	5,4	5,5	5,6
6,1	6,2	6,3	6,4	6,5	6,6

b. Sample Space Assumed by the Student

1,1	1,2	1,3	1,4	1,5	1,6
	2,2	2,3	2,4	2,5	2,6
		3,3	3,4	3,5	3,6
			4,4	4,5	4,6
				5,5	5,6
					6,6

4.2 Conditional Probability and the Fallacy of the Time Axis

Falk (1979; 1999) proposed the following problem to eighty-eight university students and found that while students easily answered part (a), they were confused about part (b).

Problem

An urn contains two white balls and two red balls. We pick up two balls at random, one after the other without replacement. (a) What is the probability that the second ball is red, given that the first ball is also red? (b) What is the probability that the first ball is red, given that the second ball is also red?

Students typically argued that, because the second ball had not been drawn at the same time as the first ball, the result of the second draw could

not influence the first. Hence, the students claimed that the probability in part (b) is 1/2.

Falk suggested that these students confused conditional and causal reasoning and also demonstrated the *fallacy of the time axis*. That is, they thought that one event could not condition another event that occurs before it. This is false reasoning, because even though there is no causal relation from the second event to the first one, the information in the problem that the second ball is red has reduced the sample space for the first drawing. In essence, there are now just one red ball and two white balls for the first drawing. Hence, P ($B1$ is red/ $B2$ is red) $=1/3$. Our hypothesis is that there is a semiotic conflict in the use we make of the words *dependent/ independent* in statistics and in other areas of science and mathematics, where this word has a causal meaning that students bring to the study of statistics (Díaz and de la Fuente 2006).

4.3 Ruling out Semiotic Conflict Does Not Always Solve Probabilistic Misconceptions

Of course the conjecture that the above difficulties might be explained by the participants' semiotic conflicts in interpreting the tasks should be empirically tested. On the other hand, there are many other probabilistic misconceptions and errors which seem to defy semantic explanations. One such example is the conjunction fallacy. In related research, many students and professionals were presented with different variations of the following problem (Kahneman and Tversky 1982).

Problem

Linda is 31 years old, single, outspoken, and very bright. She majored in philosophy. As a student, she was deeply involved with issues of discrimination and social justice, and also participated in antinuclear demonstrations. The task is to rank various statements by their probabilities, including these two:

(a) Linda is a bank teller

(b) Linda is a bank teller and is active in the feminist movement

Many students and professionals ranked response (b) as more probable than (a). This judgment apparently violates the probability rules, since sentence (a) refers to the probability of a single event $P(B)$, while sentence (b) refers to the probability of the compound event $P(B \text{ and } F)$, which at best is equal to $P(B)$ assuming $P(F)=1$.

Some researchers organized experiments to ensure that participants were not misled into interpreting (a) as "Linda is a bank teller and not active in the feminist movement" and did not therefore assign an incorrect meaning to the term probability (Sides, Osherson, Bonini, and Viale 2002). In spite of these precautions, conjunction fallacies were still very frequent. Attempts to explain the fallacy in terms of misinterpreting the word *and* in sentence (b) to refer to the conjunction of two events (even though the word possesses semantic and pragmatic properties that are foreign to conjunction) were still unsuccessful (Tentori, Bonini, and Oshershon 2004).

5. IMPLICATIONS FOR TEACHING PROBABILITY

This discussion shows the multifaceted meanings of probability and suggests that teaching cannot be limited to any one of these different meanings because they are all dialectically and experientially intertwined. Probability can be viewed as an a priori mathematical degree of uncertainty, evidence supported by data, a propensity, a logical relation, a personal degree of belief, or as a mathematical model that helps us understand reality.

The controversies that historically emerged about the meaning of probability have also influenced teaching. Before 1970, the classical view of probability based on combinatorial calculus dominated the secondary school curriculum. Since combinatorial reasoning is difficult, students often found this approach to be very hard; moreover, the applications of probability in different sciences were hidden. Probability was considered by many secondary school teachers as a subsidiary part of mathematics, since it only dealt with games of chance. In other cases it was considered to be only another application of set theory (Henry 1997; Parzysz 2003).

With increasing computer development, there is growing interest in the experimental introduction of probability as a limit of the stabilized frequency. Simulation and experiments help students solve some paradoxes which appear even in when teaching simple probability problems. A pure experimental approach, however, is not sufficient in the teaching of probability. Even when a simulation can help to find a solution to a probability problem arising in a real world situation, it cannot prove that this is the most relevant solution, because the solution will depend on the hypotheses and the theoretical setting on which the computer model is built. A genuine knowledge of probability can be achieved only through the study of some formal probability theory, and the acquisition of such theory should be gradual and supported by the students' stochastic experience.

On the other hand, much more research is needed to clarify the fundamental components of the meaning of probability (and in general of every specific mathematical concept) as well as the adequate level of abstraction in which each component should be taught, since students might have difficulties in any or all the different components of the meaning of a concept. We also emphasize the need to take into account students' semiotic activity when solving mathematical problems or carrying out mathematical activity, in order to help them overcome errors and difficulties that might be explained in terms of semiotic conflicts. Even when semiotic conflicts are not the only explanation for students' difficulties in probability, the teacher should be conscious of the existence of such conflicts and of the high semiotic complexity of mathematical work.

ACKNOWLEDGEMENT

This work was supported by the projects: SEJ2004-00789, MEC, Madrid and FEDER and grant AP2003-5130, MEC, Madrid, Spain.

REFERENCES

Batanero, C., Henry, M., and Parzysz, B. (2005). The nature of chance and probability. In G. A. Jones (Ed.) *Exploring probability in school: Challenges for teaching and learning* (15-37). New York: Springer.

Bellhouse, D. R. (2000). De Vetula: a medieval manuscript containing probability calculations. *International Statistical Review,* 68 (2), 123-136.

Bernoulli, Jacques (1987). *Ars Conjectandi- 4ème partie* (N. Meunier, Trans.) Rouen: IREM. (Original work published in 1713).

Biehler, R. (1997). Software for learning and doing statistics. *International Statistics Review,* 167-189.

Borovcnik, M., and Peard, R. (1996). Probability. In A. J. Bishop, K. Clements, C. Keitel, J. Kilpatrick, and C. Laborde (Eds.) *International Handbook in Mathematics Education* (Part 1, 239-288). Dordrecht: Kluwer.

Cabriá, S. (1992). *Filosofía de la probabilidad* (Philosophy of probability). Valencia: Tirant le Blanc.

Callaert, H. (2003). In search of the specificity and the identifiability of stochastic thinking and reasoning. *Proceedings of CERME III,* Bellaria, Italy, 2003. On line: http://www.dm.unipi.it/~didattica/CERME3/proceedings/Groups/TG5/TG5_list3.html.

Cardano, G. (1961). *The book on games of chances.* New York: Holt, Rinehart and Winston. (Original work Liber de Ludo Aleae; included in Opera Omnia, vol.1, published in 1663).

Carnap, R. (1950). *Logical foundations of probability,* Chicago: University of Chicago Press.

Díaz, C., and de la Fuente, I. (2006). Assessing psychology students' difficulties with conditional probability and bayesian reasoning. In A. Rossman y B. Chance (Eds.)

Proceedings of the Seventh International Conference on Teaching Statistics. Salvador (Bahia), Brazil: International Association for Statistical Education. CD ROM.

Eco, U. (1979). *Tratado de semiótica general* (General semiotics). Barcelona: Lumen.

Falk, R. (1979). Revision of probability and the time axis. In *Proceedings of the third International Conference for the Psychology of Mathematics Education* (64-66). Warwick, U. K: Organising Committee.

Falk, R. (1989). Inference under uncertainty via conditional probabilities. In R. Morris (Ed.) *Studies in mathematics education: The teaching of statistics*, 7, 175-184, Paris: UNESCO

Finetti, de B. (1937). La prévision: ses lois logiques, ses sources subjectives. *Annales de l'Institute Henri Poincaré*, 7, 1-68.

Gal, I. (2005). Democratic access to probability: Issues of probability literacy. In G. A. Jones (Ed.), *Exploring probability in school: Challenges for teaching and learning.* (39-63). New York: Springer.

Gillies, D. (2000). Varieties of propensities. *British Journal of Philosophy of Science*, 51, 807-835.

Godino, J. D. (2002). Un enfoque ontológico y semiótico de la cognición matemática. (A semiotic and ontological approach to mathematical cognition). *Recherches en Didactiques des Mathematiques,* 22, 2-3.

Godino, J. D., and Batanero, C. (1994). Significado personal e institucional de los objetos matemáticos (Institutional and personal meaning of mathematical objects). *Recherches en Didactiques des Mathématiques,* 14 (3), 325-355.

Godino, J. D., and Batanero, C. (1998) Clarifying the meaning of mathematical objects as a priority area of research in Mathematics Education. In A. Sierpinska, and J. Kilpatrick (Eds.) *Mathematics Education as a Research Domain: A Search for Identity* (177-195). Dordrecht: Kluwer.

Godino, J. D., and Batanero, C. (1999). The meanings of mathematical objects as análisis units for didactics of mathematics. In I. Schwank (Ed.) *Proceedings of the First Conference of the European Society for Research in Mathematics Education* (Vol. 2, 232-244). Osnabrück: Forschungsinstitut für Mathematikdidaktik.Hacking, I. (1965). *The logic of statistical inference*. Cambridge: Cambridge University Press.

Godino, J. D., Batanero, C., and Roa, R. (2005). An onto-semiotic analysis of combinatorial problems and the solving processes by university students. *Educational Studies in Mathematics,* 60 (1), 3-36.

Godino, J. D., Contreras, A., and Font, V. (In press). Análisis de procesos de instrucción basado en el enfoque ontológico-semiótico de la cognición matemática (Analysing instruction processes with base on the onto-semiotic approach to mathematical cognition). *Recherches en Didactiques des Mathematiques.*

Hacking, I. (1975). *The emergence of probability.* Cambridge, MA: Cambridge University Press.

Henry, M. (1997). L'enseignement des statistiques et des probabilities [Teaching of statistics and probability]. In P. Legrand (Coord.), *Profession enseignant: Les maths en collège et en lycée* (254-273). Paris: Hachette-Éducation,

Huygens, C. (1998). *L'art de conjecturer* [The art of conjecturing]. (J. Peyroux, Trans.). Paris: A. Blanchard, (Original work Ratiociniis in aleae ludo, published 1657).

Kahneman, D., and Tversky, A. (1982). Judgments of and by representativeness. In D. Kahhenam, P. Slovic and A. Tversky (Eds.) *Judgment under uncertainty: Heuristics and biases.* Cambridge: Cambridge University Press.

Keynes, J. M. (1921). *A treatise on probability.* New York: Macmillan.

Laplace P. S. (1985). *Ensayo filosófico sobre las probabilidades* (Philosophical essay on probability). Madrid: Alianza Editorial (Original work published in 1814).

M.E.C. (1992). *Decretos de Enseñanza Secundaria Obligatoria* (Secondary school guidelines). Madrid: Ministerio de Educación y Ciencia.

Mellor, D. H. (1971). *The matter of chance*. Cambridge. Cambridge University Press.

Mises, R. von (1952). *Probabilidad, estadística y verdad* [Probability, statistics and truth]. Madrid: Espasa Calpe (Original work published 1928).

Moivre, A. de (1967). *The doctrine of chances* (3rd ed.). New York: Chelsea Publishing (Original work published in 1718).

N.C.T.M. (2000). *Principles and standards for school mathematics.* Reston, VA; N.C.T.M. http://standards.nctm.org/

Parzysz, B. (2003). L'Enseignement de la statistique et des probabilités en France: Évolution au cours d'une carriére d'enseignant (période 1965-2002) (Teaching of statistics and probability in France. Evolution along a teacher's professional work (period 1965-2002). In B. Chaput (Coord), *Probabilités au lycée.* Paris: Commission Inter-Irem Statistique et Probabilités.

Pascal, B. (1963a). Correspondance avec Fermat (Correspondence with Fermat). In *Oeuvres Complètes* (43-49). París: ed. Seuil. (Original letter in 1654).

Pascal, B. (1963b). Traité du triangle arithmétique et traités connexes (Traitise of arithmetical triangle and related treatises). In *Oeuvres Complètes* (50-100). París: ed. Seuil. (Original letter in 1654).

Pratt, D. (2000). Making sense of the total of two dice. *Journal for Research in Mathematics Education,* 31, 602-625.

Peirce, C. S. (1932). Notes on the doctrine of chances. In *Collected papers of C. S. Peirce* (Vol 2. 404-414). Havard University Press. (Original work published in 1910).

Popper, K. R. (1957). The propensity interpretation of the calculus of probability and the quantum theory. In S. Körner (Ed.) *Observation and interpretation The Colston Papers* (Vol. 9, 65-70). University of Bristol.

Popper, K. R. (1959). The propensity Interpretation of probability. *British Journal of the Philosophy of Science,* 10, 25-42.

Ramsey, F. P. (1926). Truth and probability. In F. P. Ramsey (1931), *The Foundations of Mathematics and other Logical Essays* (Ch. VII, 156-198), edited by R.B. Braithwaite, London: Kegan, Paul, Trench, Trubner and Co.

Sides, A., Osherson, D., Bonini, N., and Viale, R. (2002). On the reality of the conjunction fallacy. *Memory and Cognition,* 30(2), 191-198.

Székely, G. J. (1986). *Paradoxes in probability theory and mathematical statistics.* Dordrecht, the Netherlands: Reidel.

Tentory, K., Bonini, N., and Osherson, D. (2004). The conjunction fallacy: a misunderstanding about conjunction? *Cognitive Science,* 28, 467-477.

Truran, K. M. (1995). Animism: A view of probability behaviour. In B. Atweh and S. Flavel (Eds.) *Proceedings of the Eighteenth Annual Conference of the Mathematics Education Group of Australasia (MERGA)* (537-541). Darwin, N. T.: MERGA.

Vergnaud, G. (1982). Cognitive and developmental psychology and research in mathematics education: some theoretical and methodological issues. *For the Learning of Mathematics* 3 (2), 31-41.

Vergnaud, G. (1990). La théorie des champs conceptuels (Theory of conceptual fields). *Recherches en Didactiques des Mathématiques,* 10 (2-3), 133-170.

Vygotski, L. S. (1934). *Pensamiento y lenguaje.* (Thought and language). Madrid: Visor, 1993.

INTERLUDE 6

The transition from Carmen Batanero and Carmen Díaz's paper to Maria Meletiou-Mavrotheris's paper seems quite obvious: probability and stochastics are as close as you want. So basically they are dealing with the same subject, are they not? Well, it depends. If we look at the topic from a *pure* mathematical point of view, yes of course; but if we put the topic into another context, such as the classroom, then things become quite different.

As it happens, both authors agree that there are serious problems as to the *translation* of the pure mathematics into classroom understanding and the insights required for specific applications. Then where do we find the differences? Quite simply, in their diagnosis. Looking for answers to the question "Why do students perform so poorly?", different answers are given by Batanero and Díaz and by Meletiou-Mavrotheris.

The former authors focused on the problem of the multiplicity of meanings, the latter author looks at the sharp contrast between the formal notion of probability and stochasticity and the corresponding everyday notions (i.e., the way we humans in our everyday lives think, talk, and reason about chance events, probabilities, and the like). In a way, it is close to a miracle that so many mathematicians and philosophers are profoundly shocked when they learn from the cognitive psychologists that everyday people reason differently from their idealized formal alter egos. Unfortunately (or fortunately?), we no longer recognize our alter egos because they are not even relatives of ours, let alone a variation of ourselves. But don't we have a problem now? If the answers of both authors are different, does that not mean that a choice must be made, that someone is right and someone is wrong? Perhaps, if the question is "What is 2 + 2?," and two answers "4" and "5" were given (although in this case too, perhaps not, as it could very well be that both answers are wrong, the right answer being: "an addition.") What needs to be done is to integrate the ideas in both papers into a more encompassing (semiotic?) theory that takes into account the curious kind of being we are, a being that generates concepts and then *discovers* it fails to understand them. Are intuitions going berserk?

THE FORMALIST MATHEMATICAL TRADITION AS AN OBSTACLE TO STOCHASTICAL REASONING

Maria Meletiou-Mavrotheris
Cyprus College, Cyprus and Central Michigan University, USA

Abstract: This paper argues that the persistence of students' difficulties in reasoning about the stochastic, despite significant reform efforts in statistics education, might be the result of the continuing impact of the formalist mathematical tradition. It first provides an overview of the literature on the formalist view of mathematics and its impact on statistics curricula and instructional approaches. It then reconsiders some well-known empirical findings on students' understanding of statistics, and forms some hypotheses regarding the link between student difficulties and mathematical formalism. Subsequently, it suggests the pedagogical and curricular changes that ought to take place in order to move away from the formalist mathematical tradition and to improve people's impoverished probabilistic and statistical reasoning. Finally, it briefly discusses possible research directions for the formal study of the effects of the formalist tradition on statistics education.

Key words: Formal reasoning, everyday reasoning, statistics instruction, formalist view, intuitions

1. INTRODUCTION

New values and competencies are necessary for survival and prosperity in the rapidly changing world, where technological innovations have made redundant many skills of the past (Ghosh 1997). The Lisbon European Council of March 23-24, 2000, placed the development of a knowledge-based society at the top of the Union's policy agenda, considering it to be the key to the long-term competitiveness and the personal aspirations of its citizens. Statistics education has a crucial role to play in this regard. The pressure for democratization of mathematics education and the shift of

mathematical studies towards a more "utilitarian approach" (Moore 1997), have opened up a larger place for statistics, which now, at the post-secondary level, is probably studied by more students than any other topic (Philips 1999). The development of a statistically literate society is a key factor in achieving the objective of an educated citizenry. However, the research literature in the area of stochastics education indicates that most people tend to think deterministically and to have weak intuitions about the stochastic. Students' difficulties persist despite the significant reform efforts that have led to wide-scale incorporation of technology and interesting activities in the statistics classroom.

Wilensky (1993) has claimed that the failure in developing sound probabilistic intuitions is similar to other failures in mathematical understanding and is the result of deficient learning environments and reliance on "brittle formal methods." It is, in my opinion, these same reasons that also cause the neglect of a number of other statistical concepts observed both in the curriculum and in the research literature. It is argued here that deep-rooted beliefs about the nature of mathematics are imported into statistics, affecting instructional approaches and curricula and acting as a barrier to the kind of instruction that would provide students with the skills necessary to recognize and intelligently deal with uncertainty and variability.

In this paper, I give an overview of the literature on the formalist view of mathematics and its impact on statistics instruction and learning. While little empirical work has focused on examining the impact that formalism may have on student understanding, I use here examples from a number of studies in statistics that discuss student understanding in general and I illustrate how student difficulties may be related to mathematical formalism. Next, I provide suggestions about the pedagogical and curricular changes that ought to take place in order for mathematics and statistics education to break away from the formalist tradition and ensure that all students develop their ability to reason effectively about the stochastic. Finally, I discuss possible directions for research on this important topic.

2. FORMALIST VS. RELATIVISTIC VIEW OF MATHEMATICS

In recent years, the *formalist* tradition in mathematics and science has come under attack and a second agenda, which views mathematics as a meaning-making activity of a society of practitioners (Wilensky 1993), has begun to emerge. The emergence of the new paradigm has been the result of developments in the history and philosophy of science, which have caused a general shift in the last thirty years in virtually every social science and field

of humanities away from rationalistic, linear ways of thinking. In the social sciences several critics have attacked formalist tradition in mathematics and science. Hermeneutic critics (Packer and Addison 1989 in Wilensky 1993) have criticized it for its detachment from context, its foundation on axioms and principles rather than practical understanding, and its formal, syntactic reconstruction of competence. Feminists have criticized it for alienating a large number of people, especially women (Gilligan 1986). Sociologists such as Latour (1987) have maintained that science can only be understood through its practice.

Kuhn (1962) has argued that the progress of science, rather than being linear and hypothetico-deductive as claimed by logical positivists, is made possible through revolution. New theories are not incremental modifications of existing ones but ideas that posit basic entities of the world, which are fundamentally incompatible with old theories. An anomaly occurs, which is an event that cannot be explained by the existing theory. In effect, the theory/paradigm is disproved. A whole new paradigm takes over, which explains everything the old paradigm explained and also explains the anomaly (Wiesman and Wotring 1997). Pólya (1962) and Lakatos (1976) have argued that, "by placing such a strong emphasis on mathematical verification and the justification of mathematical theorems after their referent terms have been fixed, mathematics literature has robbed our mathematics of its basic life" (Wilensky 1993, 34). Mathematics for Lakatos is a human enterprise and advances happen through the negotiation of meaning among a community of practitioners. It is not given in advance but is constructed through the practices, needs, and applications of this community of practitioners. Proofs are not developed in a linear way, but follow "the *zig-zag* path of example, conjecture, counter-example, revised conjecture or revised definition of the terms referred to in the conjecture" (Wilensky 1993, 34). In the new paradigm, the history of mathematics takes an important role. Its examination reveals that "mathematics is messy and not the clean picture we see in textbooks and proofs," that "the path to our current mathematical conceptions was filled with argument, negotiations, multiple and competing representations" (Wilensky 1993, 17).

In response to criticisms following research findings, and reports of the 1970s and early 1980s exposing students' impoverished understanding of mathematics and science, leaders and professional organizations in mathematics education are now finally promoting a relativistic view of mathematics (Confrey 1980; Nickson 1981). They have come to believe that current teaching approaches are deficient in that they do not give students the chance to encounter different perspectives on the nature and uses of mathematics. Reformers argue that the culture of the mathematics classroom should change. Mathematics should be presented as open to discussion and

investigation, as a socially constructed discipline which, even at the classroom level, "is not held to be exempt from interpretations that require 'reconsideration, revision and refinement'" (Nickson 1992, 104). The emphasis "should not be on mirroring some unknowable reality, but in solving problems in ways that are increasingly useful to one's experience" (Confrey 1991, 136). The teacher should encourage discussion, and allow students to generate and test their own theories.

Wilson, Teslow, and Osman-Jouchoux (1995) warn us that although recent models of cognition are clearly challenging our traditional notions of learning and teaching, changing long-held beliefs and attitudes towards mathematics is proving to be quite difficult. There is an enormous gap between the mechanistic-instrumental portrayal of the nature of mathematics and the more realistic-fundamental view that reformers try to advance. The formalist tradition has been around for too long and runs deeply into people's veins. For people raised in this tradition, it is very difficult to accept the fallabilist nature of mathematics.

3. IMPACT OF FORMALIST VIEW ON STATISTICS EDUCATION

In the statistics domain, there has already been a move towards modernizing statistics education and a general acknowledgment that learning occurs most effectively when students engage in authentic activities. Statistics educators are pushing for courses that put more emphasis "on data collection, understanding and modeling variation, graphical display of data, design of experiments and surveys, problem solving, and process improvements, and less emphasis on mathematical and probabilistic concepts" (Ballman 1997). This has caused a movement away from statistical content emphasizing the abstract and the memorization of a list of formulas and procedures, toward content emphasizing exploratory data analysis. Although many students are still being taught in traditional classrooms, there is already a large number of statistics instructors who have adopted alternative approaches to their teaching and many statistics classrooms are experiencing wide incorporation of technology. But, as Hawkins (1997a) points out, for reform efforts to be successful, it is also necessary to change attitudes and expectations about statistical education. Changing long-held beliefs and attitudes towards statistics is proving to be quite difficult (Wilensky, 1993). Many people still view statistics as "a branch of the older discipline of mathematics that takes its place alongside analysis, calculus, number theory, topology, and so on" (Glencross and Binyavanga 1997, 303). This affects statistics instruction and hampers the reform efforts.

The linear and hierarchical approach adopted by statistical courses and syllabuses is testimony to the profound and continuing effect of the formalist mathematics culture on statistics. The structure of almost every introductory statistics course is to start with descriptive and exploratory data analysis, then move into probability, and finally go to statistical inference. It is assumed that this simplifies the process of learning by gradually leading students from more basic to more complex connections (Steinbring 1990). However, presenting statistical content as a sequenced list of curricular topics might lead to compartmentalization of knowledge and fail to communicate to students the interconnectedness of the different statistical ideas they encounter in the course.

Statistical methods were developed to help us filter out any *signals* in data from surrounding *noise*. *Signals* are the messages, the meanings we find in *explained variation*, the patterns that we have not discounted as being transient. *Noise* is the *unexplained variation* that remains after we have *removed* all patterns. Randomness and probability theory are *human constructs* created to deal with unexplained variation. We use probability theory to model and describe phenomena in the world for which no patterns can be discerned, assuming that they had been randomly generated. Thus, "what probability is can only be explained by randomness, and what randomness is can only be modeled by means of probability" (Steinbring 1990, 4). Stochastical knowledge is created as "a relational form or linkage mechanism between formal, calculatory aspects on the one hand, and interpretative contexts on the other" (Steinbring 1990, 5). The classroom culture, however, often comes in sharp contrast with this conception of stochastical knowledge as being developed through a "self-organized" process that balances the objective aspects of a situation and the formal means employed to model and describe it.

The linear, completely elaborated, and hierarchical structure of knowledge presentation encourages the development of the chance concept as a concrete, totally clear and unambiguous generalization defined by methodological conventions. Steinbring (1990), who analyzed teaching episodes from several different classrooms in order to see how the concept of chance was introduced, found that in all instances *chance* was first introduced through performing and discussing a chance experiment. An attempt was then made to describe the experimental outcomes using a rule or a simple stochastical model. Naturally, there was always variation observed between the theoretical predictions and the empirical data. The pattern of justification for the observed variation, regardless of its size, always was that the difference between the empirical result and the theoretical prediction was produced by *chance* (Steinbring 1990). The difference between theory and experiment was thus neutralized, with chance degenerating into "a substitute

for justification, which serve[d] to deny the importance of the difference between theory and empirical facts in probability" (Steinbring 1990, 14).

The assumptions posed in the statistics classroom are often too simplistic. Although not necessarily denying underlying causal explanations in the case of chance events, a probabilistic approach views them as impractical and adopts a *blackbox* model (Biehler 1994, 10). As Biehler indicates, however, the assumption of independence is not plausible in many real world contexts: "Even coin flipping can be done in a way that independence has to be rejected in favor of serial correlation, and physical theories can be developed to explain some aspects of coin flipping" (Biehler 1994, 10). Borovcnik and Peard (1996) warn that instruction has traditionally underrated the complexity and dangers of using pseudo-real examples that conflict with students' emotions or with their common sense.

The over-emphasis of the traditional mathematics curriculum on determinism and its "orientation towards exact numbers" (Biehler 1997, 187) affects statistics instruction, becoming an obstacle for the adequate judgment of stochastic settings. The law of large numbers is often presented as a canon in the statistics classroom, giving students the false impression that the stabilization of the relative frequency of repeated sampling to the ideal value is *guaranteed.* Similarly, traditional instruction leaves students with the impression that a larger random sample *guarantees* a more representative sample. There is a deterministic mindset and an over-reliance on rules and theorems, forgetting that we are dealing with uncertainty; and the variability accompanying all finite statistical processes implies that a sample is almost never totally representative of the population from which it was selected. People have a hard time distinguishing between the real world problem and the statistical model. At one extreme are many people who use statistical methods for solving real world problems in the same way that they would use an artificial mathematics problem coming out of a textbook. At the other extreme, we find people who distrust statistics completely due to the fact that, unlike mathematics, it deals with uncertainty. Both of these two extreme attitudes suggest inadequate understanding of statistics as a decision-support system (Biehler 1997).

4. FORMALISM AND STUDENT DIFFICULTIES WITH STATISTICAL REASONING

Despite the criticisms regarding the impact of formalism on statistics education, little empirical work has been done towards the better understanding of the difficulties students face that may or may not relate to formalism. Therefore, little is known regarding the details of how

misconceptions are formed and how they may be prevented. I hereby attempt to touch upon this issue through a reconsideration of some well-known empirical findings on students' understanding of statistics. I compiled a list of difficulties students face, as documented in the literature, and I formed some hypotheses regarding the link between these difficulties and the formalist tradition.

4.1 Over-reliance on Sample Representativeness

Statistical reasoning follows from two notions which, when seen from a deterministic framework, seem antithetical—*sample representativeness* and *sample variability*. Whereas sample representativeness is "the idea that a sample taken from a population will often have characteristics identical to those of its parent population," sample variability is "the contrasting idea that samples from a single population are not all the same and thus do not all match the population" (Rubin, Bruce, and Tenney 1990, 3). Due to sample representativeness we can put bounds on the value of a characteristic of the population; due to sampling variability however, we never know exactly what that value is (Rubin et al. 1990). Balancing these two ideas lies at the heart of statistical inference. Although recognizing that random selection always leads to variation, however, most students tend to underestimate the effect of sampling variability and, over-relying on sample representativeness, they search for patterns in the data with a certainty that such patterns exist – an outcome of their training in formalist tradition. Indeed, mathematics teaching with roots in formalist tradition often encourages this searching for patterns. In statistics however, when reasoning in terms of patterns, students often fail to conceptualize the chance variation involved in those patterns and hence often exaggerate the information provided. Students view random fluctuations in data as causal and proceed to develop deterministic explanations.

4.2 Neglect of Variation

A common finding in statistics studies is a neglect of variation (e.g., Meletiou 2000; Meletiou-Mavrotheris and Lee C. 2002; 2003). Most people have a limited ability to cope with uncertainty and variation in the real world. They tend to think deterministically and to have difficulties in differentiating between chance variation in the data and variation due to some form of underlying causality.

Joiner and Gaudard (1990) consider awareness of variation and how it affects processes to be one of the main determinants of success in business management. Hoerl, Hahn, and Doganaksoy (1997) point to the gross

inefficiencies that occur in industry because managers and technical personnel have a deterministic mindset and lack awareness of variation, despite the fact that many of them have had substantive formal statistical training: "They expect mass balances to match exactly, or actual financial figures to exactly equal budget. Any *variance* from budget must be explained" (Hoerl et al. 1997, 149). A partial explanation Hoerl et al. (1997) see for the shortcomings in statistics education goes back to the math vs. statistics issue:

> Perhaps because of the over-emphasis on mathematics, statisticians seem uncomfortable with statistical concepts which cannot be derived or proven mathematically. The omnipresence of variation is admitted, but often not clearly explained. (Simulating a histogram on the computer does not teach business people how to interpret a financial report.) (Hoerl et al. 1997, 150)

The results of a study conducted by Shaughnessy, Watson, Moritz, and Reading (1999) to investigate elementary and high school students' understanding of variability indicated a steady growth across grades on center criteria, but no clear corresponding improvement on spread criteria. Indeed, most students treat the task of finding the center of a dataset as a typical mathematics task requiring the application of a simple formula for its solution. Ignoring the possibility of outliers, they rush into adding up all the numbers and dividing by the number of data values. Further, when asked to compare group means, students tend to focus exclusively on the difference in averages and to believe that any difference in means is significant.

I hypothesize that this over-emphasis on center criteria and tendency to underestimate the effect of variation in real world settings is related to the emphasis of the traditional mathematics curriculum on determinism and its orientation towards exact numbers. Since centers are often used to predict what *will* happen in the future, or to compare two different groups, the incorporation of variation into the prediction would confound people's ability to make clean predictions or comparisons (Shaughnessy 1997). The formalist tradition prepares students to search for the one and only correct answer to a problem—a condition that can easily be satisfied by finding measures of center such as the mean and the median. Variation, though, rarely involves a *clean* numerical response. Standard deviation, the measure of variation on which statistics instruction over-relies, is computationally messy and difficult for both teachers and curriculum developers to recommend to students as a good choice for measuring spread (Shaughnessy, 1997).

4.3 Local Representativeness Heuristic - Perceiving Patterns in Random Data

The research literature has identified a series of heuristics often subject to bias that humans develop in an effort to rationalize stochastic events. These heuristics indicate people's limited understanding of randomness, as well as their tendency to reason deterministically and to develop causal explanations for random fluctuations in the data. Well-documented in the research heuristic is *local representativeness*, the phenomenon in which "people believe that a sequence of events generated stochastically will represent the essential characteristics of that process, even when the sequence is quite short" (Pratt 1998, 37). For example, when tossing coins, people consider it less likely to obtain HHHTTT or HHHHTH than to obtain HTHTTH, because HTHTTH seems to better represent the two possible outcomes. Similarly, the fallacy of the gambler who, after a long sequence of red outcomes, expects the next outcome to be black is, for Kahneman and Tversky (1973), the consequence of employing the local representativeness heuristic and perceiving a pattern in random data. The gambler's fallacy is also called the "law of averages" as it describes people's tendency to believe that things should balance out to better represent the population distribution. This is the same idea as that which Shaughnessy (1992) calls active balancing strategy. For him, an active balancer is the person who, when given the problem "The average SAT score for all high school students in a district is known to be 400. You pick a random sample of 10 students. The first student you pick had an SAT of 250. What would you expect the average of the other scores to be?" (Shaughnessy 1992, 477), would predict the average of the remaining 9 scores to be higher than 400, in order to make up for the "strangely" low score.

Conventional instruction often fails to establish enough links between the learners' primary intuitions about the stochastic and "the clear cut codified theory of the mathematics" (Borovcnik and Bentz 1991; in Pfannkuch and Brown 1996). Students coming to the statistics class have already experienced the highly fluctuating and irregular pattern of random phenomena such as the occurrence of *Heads* and *Tails* in sequential coin tosses. The theoretical statement that *P (Head on next toss)=1/2*, which describes the relative limiting frequency of an event, seems to students to be in sharp contrast to the intuitively felt inability to make specific predictions of this outcome (Borovcnik 1990). Even if students understand probabilistic theory, they often fall back into the trap of causal thinking. In their urge to overcome uncertainty, to order the chaos, they might attempt to search for logical patterns, to develop *different mathematical theories* and causal links. Such an approach "is highly interwoven with magic belief and astrology (the

law of series, a change is overdue, etc.) and the search for the signs to detect this early enough" (Borovcnik 1990, 8); it leads to the development of heuristics such as the local representativeness heuristic.

Fischbein (1975) notes that although probabilistic intuitions exist from a very early age they are suppressed by schooling. In order to make his point, he reports on a study where participants were asked to predict the outcomes of a repetitive series of stochastic trials, and where even young children were able to make sound predictions based on the relative frequencies of the different outcomes. The reason that the intuition of chance remains excluded from intellectual development is the emphasis of school mathematics on causality and determinism and its sole focus on deductive reasoning. Due to the lack of nourishment of probabilistic intuitions, or (as I argue in this article) due to the training in formalist traditions that discourage non-deterministic reasoning, learners develop a series of heuristics often subject to bias, in an effort to rationalize stochastic events.

4.4 The Outcome Orientation

This heuristic also describes students' tendency to interpret in deterministic terms phenomena that are actually stochastic (Konold 1989). Lacking awareness of the stochastic dimension of such phenomena, students often make predictions based solely on causal factors (Pratt 1998). Influenced by tasks posed in mathematics classrooms to which there is always a right answer, students tend to deal with uncertainty by predicting what the next outcome will be and then by evaluating the prediction as either right or wrong. A probability of 50 percent is often assigned when no sensible prediction is possible. The information that there is a 50 percent chance of rain tomorrow sounds totally useless, a probability of 30 percent implies that there is no possibility of rain, whereas a probability of 70 percent means that it will definitely rain.

4.5 Disconnection from Context – Good Mathematics Is "Pure"

In the statistics classroom, concepts related to probability are most often taught through standard probability tasks such as throwing dice and tossing coins. This norm of using pseudo-real examples, borrowed from mathematics instruction, does not serve students well. As the research literature indicates, the ability to solve problems involving random devices does not transfer very effectively to more applied problems (Garfield and delMas 1990; Pfannkuch and Brown 1996; Meletiou-Mavrotheris and Lee C.

2002; 2003). People's understanding of probability is more limited in real world contexts than in the contrived context of standard probability tasks.

According to Nisbett, Krantz, Jepson, and Kunda (1983), dealing with standard random devices makes easier the recognition of the operation of chance factors than dealing with social events. Random devices have an obvious sample space and the repeatability of trials can be easily imagined. By contrast, the random nature of social events is often not as explicit and the sample space not well understood, since use of a real world context increases the likelihood of prior beliefs and knowledge about the issues under investigation. Students comfortable thinking probabilistically when dealing with standard probability tasks, seem *oblivious* to probabilistic thinking for problems posed in real world settings. Although, for example, students might be aware of the dangers involved when drawing conclusions from small samples, for problems posed in realworld contexts they often ignore these dangers and do not hesitate to use small samples as a basis for inferences, erring thus towards the deterministic side (Pfannkuch and Brown 1996; Meletiou-Mavrotheris and Lee C. 2002; 2003).

5. MOVING AWAY FROM THE FORMALIST MATHEMATICAL TRADITION: IMPLICATIONS FOR INSTRUCTION

In this paper, I have reconsidered some well-known empirical findings on students' understanding of statistics, and asserted that student difficulties might stem from training in formalist mathematics traditions. As a consequence of their prior experience with traditional mathematics curricula and classroom settings that discourage non-deterministic reasoning, students have not developed adequate intuitions about the stochastic. Statistics instruction itself, also influenced by the formalist mathematics tradition, fails to build bridges between students' intuitions and statistical reasoning. Although notions such as randomness and variation have a nature very much dependent on context and lend themselves especially well to the new perspective of mathematical concepts as social constructs, they are typically presented in the classroom as rigidly established bodies of mathematical knowledge without any reference to real world context. As a result, instruction fails to convey to students the relationship between the knowledge they acquire in the statistics classroom and its uses in the real world.

Society's expanding use of data for prediction and decisionmaking in all domains of life makes data literacy increasingly important in our modern society. Recognizing uncertainty as a characteristic of reality, and behaving

intelligently within it, now form a fundamental part of the intellectual development of the individual (Azcarate and Cardnoso 1994). Thus we could no longer accept instructional practices that lead to the strong inclination of interpreting the world in deterministic ways currently observed among most adults. It becomes a priority for mathematics and statistics education to break away from the formalist tradition in order to ensure that all students develop their stochastical reasoning (National Council of Teachers of Mathematics 2000).

5.1 Redefining Statistics Instruction

Current practices in statistics education have evolved from a background quite different from today's needs and possibilities. Hawkins (1997b) argues that nothing should be taken for granted. Reform efforts "must have the momentum and energy to challenge even the most fundamental and widely-held ideas about statistical education, and the ways in which these are currently manifested" (Hawkins 1997b, 142). Technological advances and the forces of democratization demand fundamental pedagogical as well as curricular changes that would make statistical reasoning more accessible to all students.

The review of the research literature and insights gained from personal research lead us to conclude that in order to build connections between formal mathematical expressions of the stochastic and everyday informal intuitions, we have to redefine statistics instruction based on the following principles:

5.1.1 Interconnectedness of Statistical Ideas

The current practice in almost every introductory course of presenting statistics content as a sequenced list of curricular topics fails to communicate to students the interconnectedness of the different statistical ideas they encounter in the course. The assumption that by building concepts "separately but directly", students would eventually have an array of statistical ideas at their disposal (Lachance and Confrey 1996), is a very simplistic view of the development of human understanding. Learning about a statistical concept without exploring its connection to the other main statistical constructs can only lead to weak and narrow understandings. As Lachance and Confrey (1996) assert, "the best route between two points is not always a straight line" (Lachance and Confrey 1996, 23). Instruction should instead provide an interconnected path which would better encourage students to follow their own, unique *nonlinear* developmental paths from *smaller* to *larger* ideas (Lachance and Confrey 1996). Following such a path,

rather than a compartmentalized statistics curriculum, would lead students to stronger and deeper understandings.

5.1.2 Complementarity of Theory and Experience

In the increasingly many sectors of society relying on data, the purpose of a statistical investigation is to help inform decisions and actions by expanding the existing body of context knowledge about the situation under study. Therefore the ultimate goal of statistical investigation is "learning in the context sphere" (Wild and Pfannkuch 1999). This means much more than collecting new information or understanding different statistical ideas in isolation. Statistical reasoning necessitates a complementarity of theory and experience. It is not a separable entity but a synthesis of statistical knowledge, context knowledge, and the information in the data in order to produce implications, insights, and conjectures (Wild and Pfannkuch 1999).

The arid, context-free landscape on which so many examples used in statistics teaching are built ensures that large numbers of students never ever see, let alone engage in, statistical reasoning. If the statistics classroom is to be an authentic model of the statistical culture, it should model realistic statistical investigations rather than teaching methods and procedures in a sequential manner and in isolation. The emphasis should be on the statistical process. The teaching of the different statistical tools should be achieved through putting students in a variety of authentic contexts where they need those tools to make sense of the situation. Students should come to view and value statistical tools as a means to describe and quantify the variation inherent in almost any realworld process.

The teaching of probability concepts should be achieved through realistic investigations and not through the use of pseudo-real examples. As already pointed out, research has proved naïve the expectation that students will transfer the understanding obtained through coins, dice, and games of chance to everyday contexts. The skills required to understand variation of outcomes in random devices are very different from the skills required to understand variation of outcomes in reallife contexts. Consequently, instruction ought to take into account the great variety of prior beliefs, conceptions, and interpretations that students bring to each situation.

The concepts of randomness and probability should be organized under a dynamical perspective rather than being presented as ready-made elements through clear-cut context-free definitions, as is currently the case in most statistics curricula. There is a need for intuitive representations to help students "see the fundamental relationship between chance and regularity, between irregular, unpatterned phenomena on the one hand, and the mathematical intentions to model and describe them in a regular and formal

way on the other" (Steinbring 1990, 3). Probability theory should not be presented as a body of clear and unambiguous mathematical generalizations free of any concrete interpretations, but as an attempt by humans to attain a degree of certainty in contexts where "it is no longer possible to advance certain predictions about future events on the basis of strictly causal linkages" (Steinbring 1990, 2).

Particular attention should be paid to the relation between probability theory and empirical outcomes. Doing stochastic experiments and simulations and evaluating experimental outcomes, provides a socially constituted teaching context that opens up two main possibilities for further development of the chance concept. The first possibility is what is typically observed in classrooms—"a narrowing reduction of the chance concept to a formalized, conventional label" (Steinbring 1990, 17). Often stark contradictions between theoretical prediction and empirical observation are justified as being the result of "chance" in the naïve sense. The second possibility is for instruction to broaden the chance concept to that of a device for controlling the underlying connection between the stochastical model and the experimental situation. Experiments and computer simulations performed in the classroom to facilitate learning of statistical concepts should be perceived as fundamental sources for the students and not simply as motivations for step-by-step teaching of the teacher's intended goals (Steinbring 1990). The self-referent epistemological structure of stochastical concepts should also be reflected in the social process of the classroom. Stochastical knowledge necessitates direct subjective decisions and interpretations; thus contradictions between theoretical arguments and empirical results pointed out by students should not overlooked.

Whenever a very rare event is observed, one ought to question the whole process, and decisions must be taken as to whether it would be necessary to modify some basic assumptions or the experimental conditions of the process (Steinbring 1990). Such an approach, of lifting chance from "a naïve intuitive concept, which is defined only negatively as non-existing regularity" (Steinbring 1990, 18), to a theoretical concept that calls for careful analysis of experimental conditions and theoretical assumptions, emphasizes the complexity of real life situations, rather than making simplistic assumptions that conflict with students' common sense. It permits the re-establishment of an appropriate balance between objective and subjective aspects of knowledge and lays solid foundations for the development of the most important stochastical concepts.

5.1.3 Building on Student Intuitions

Student thinking about the stochastic has a manifold nature. It is not only probabilistic reasoning that drives students' thinking, but also impressions, prior beliefs, and expectations. Instruction that ignores students' prior knowledge and intuitions will not take us far from the formal knowledge (itself quite shaky, as the research literature suggests) of statistical methods and procedures with no connection whatsoever to reality. Konold (1995) insists that the *forget-everything* approach probability and statistics teachers often embrace in the hope that what they convey to students will be accurately encoded does not work simply because "learning is both limited and, at the same time, made possible by prior knowledge." The only way people can cope with new information is to relate it to things they already know, since "there is no blank space in our minds within which new information can be stored so as not to *contaminate* it with existing information" (Konold 1995). One cannot overwrite students' beliefs and intuitions with more appropriate ones. Unless instruction establishes direct links between the intuitive and theoretical level, students' understanding of probabilistic concepts will remain impoverished.

Pfannkuch and Brown (1996) argue that an effect of the clash between students' intuitions and probabilistic reasoning might be that students learn to distrust their intuitions, but because they do not actually understand why they are wrong, return back to them. In order to prevent this possibility, statistics education ought to find ways to help learners build powerful connections between formal mathematical expressions of the stochastic and everyday informal intuitions (Borovcnik, 1990). In order to simplify mathematical relations and build links to students' intuitions, instruction should emphasize the use of analogies from students' everyday experience, in contexts familiar to students. Instruction should revolve around students' ways of thinking and understanding, and not around a pre-determined, rule-based curriculum. "Rather than molding and shaping students to do [statistics] a certain way, and rewarding those with the *best fit*" (Scarano and Confrey 1996, 32), classroom activities and assessments should allow for a variety of perspectives and approaches. Curricular activities should be designed to be flexible and open-ended so that they can be adapted in response to feedback received from students.

5.1.4 Balance between Stochastic and Deterministic Reasoning

Students' understanding of randomness and chance variation involves not only conceptual construction, but also beliefs about the place of chance in the world. The students' epistemological set is an important dimension of

their understanding and application of ideas related to chance variation. An epistemological set is an individual's inclination to interpret the world in relative deterministic or stochastic terms, based on their beliefs about the place of chance and variation in the world (Metz 1997). Individual beliefs about the place of chance and uncertainty in the world can affect students' ability to grasp the main ideas behind inferential statistics and their propensity to apply chance interpretative schemas. Understanding, for example, that a population proportion of ¾ is a ratio of the expected relative distribution over an infinite number of repetitions of the event, but that this ratio is only approximated across many repetitions, requires more than having constructed the concepts of randomness and chance. It also requires an inclination to interpret the situation in terms of chance and uncertainty (Metz 1997).

All individuals have both chance and deterministic interpretations within their cognitive repertoire, which one they utilize depends on many factors, including the context of the situation. Nevertheless, different individuals have varying tendencies of interpreting phenomena toward one or the other end of the stochastic/deterministic continuum (Metz 1997). As we have seen, the research literature indicates that the majority of students tend to have relatively deterministic epistemological sets. The lack of nourishment of probabilistic intuitions in mathematics classrooms, results in a clear tendency among students to err toward the side of attributing too much to deterministic causality, and a failure to appreciate the extent to which chance operates in what one experiences in the world.

The idea that "chance variation, rather than deterministic causation, explains many aspects of the world" (Moore 1990, 99) is a fundamental notion for students' effective handling of data-based curricula, as well as their adequate interpretation and prediction of patterns outside of school. It is then crucial for statistics instruction to help students develop their ability to reason more effectively about the stochastic. However, instruction should not present probabilistic thinking as an alternative to deterministic thinking, but as something to be grafted on top of students' propensity to assume deterministic explanations for fluctuations in the data. Statistical thinking should be presented as a balance between stochastic and deterministic reasoning.

Viewing as a negative quality students' intuitive tendency to come up with causal explanations for any situation they have contextual knowledge about is an attitude that will not take us far in our efforts to help students improve their intuitions of the stochastic. Since most realworld problems are arise from a desire to improve something by identifying and controlling causes, we should rather view this impulse of students to find causes for phenomena as a positive resource (Wild and Pfannkuch 1999), an important

precursor of statistical reasoning. Shaughnessy (1992) reminds us that the heuristics people use to deal with uncertainty are not necessarily bad and often give good information. We should not forget that, for example, representativeness is fundamental to the epistemology of statistical events. Representativeness is, after all, the very idea that allows us to draw conclusions about an underlying population based on a random sample. We should understand that "it is not that there is something wrong with the way our students think, just that they—and we—can carry the usefulness of heuristics too far" (Shaughnessy 1992, 479). Thus we should try to create a curriculum that builds on the strengths of students' intuitive tendency to come up with causal explanations, while at the same time helping them develop secondary intuitions which will raise their awareness of *probabilistic interpretation*, and allow them to understand how stochastic thinking is related to causal thinking.

Instead of emphasizing individual irregularity, it is more constructive for statistics instruction to promote in students a way of thinking that perceives probability distributions as models of real situations that are based on some conditions which, when changed, might lead to changes in the distribution (Biehler 1994). Such an approach, which Biehler (1994) calls "statistical determinism," is especially useful for dealing with more realistic situations. Students should understand that we could analyze causes of why an individual event (e.g., a traffic accident) took place, but also realize that we can learn something from a consideration of a system of events (e.g., aggregated data on traffic accidents) that we cannot learn if we focus only on the individual event. This complementarity of individual and system level, which is usually suppressed by instructors who call attention only to the system level, is more intuitively convincing for students and more productive. It will help students realize that the best way to search for real, non-ephemeral causes is through an in-depth study of a system and not looking only for a cause among recent changes (Wild and Pfannkuch 1999).

Students know from everyday experience that, even when studies are conducted under very similar conditions, the patterns observed in one study will never appear identical in another study. What statistics instruction should be stressing is that statistical strategies, based on probabilistic modeling, are the best way to counteract our natural tendency to view patterns even where none exist, and to distinguish between real causes and ephemeral patterns that are part of our imagination. Statistics instruction should encourage students to adopt a critical attitude whenever receiving new ideas and information, to develop "a healthy skepticism and the imagination needed for alternative explanations" (Breslow 1999, 253). This critical attitude means being constantly on the outlook for logical and factual flaws, and it includes learning to counteract the human tendency to be less

judgmental of results that agree with one's predispositions, expectations, and worldviews. It can be taught by gaining experience and seeing ways in which certain types of information turn out to be false and unsoundly based.

5.1.5 Variation as the Central Tenet of Statistics Instruction

A main reason I see for students' difficulties with the stochastic is the neglect of variability and the statistical determinism hidden in standard approaches to statistics instruction. Although variation is a critical issue throughout the statistical inquiry process, from posing a question to drawing conclusions (Pfannkuch 1997), the overemphasis of the traditional mathematics curriculum on causality and determinism (Biehler 1997) affects statistics instruction, resulting in a tendency to under-recognize the role of variation in statistical reasoning. I believe that the objectives of an introductory statistics course might be better met through topics and activities that help build sound intuitions about the characteristics of random variation and its role on statistics.

Variation is the foundation of statistical thinking, the very reason for the existence of the discipline of statistics (Shaughnessy and Ciancetta 2001). Statistical thinking is concerned with learning and decisionmaking under conditions of uncertainty. Variation is a critical source of uncertainty. The fact that all processes vary is what creates the need for statistics; and it is the need to deal with variation through measurements that provides the numerical basis for comparison that produces data (Snee 1999). We use statistical tools to analyze this data and observe the pattern that exists despite (or because of) the variation. Thus, according to Snee (1999), the elements of statistical methods are variation, data, and statistical tools. Understanding of variation and using this understanding to improve the performance of processes is the core competency; and it should be the focus of statistical education, research, and practice (Snee 1999). Understanding what data is relevant and how to construct proper methods of data collection and analysis enhances successful application of this core competency (Snee 1999).

A number of other authors have stressed the central role that variation plays in statistical reasoning and that it should also play in statistics instruction. Moore argues that the essential components of statistical reasoning are (1) recognition of *the omnipresence of variation*, of the fact that "chance variation rather than deterministic causation explains many aspects of the world" (Moore 1990, 99), and (2) familiarity with the ways in which variation is quantified and explained. Pfannkuch (1997) considers the two essential and interlinked components of statistical thinking to be: (1) the recognition of variation, including the ability to distinguish between *special cause* variation (variation that can be assigned to an identifiable source) and

common cause variation (variation that is hard to link to any particular source); and (2) the realization that a sound judgment of a situation can be made only by collecting and analyzing data.

Because the central element of statistical thinking is variation, statistics instruction should aim at providing students with the skills necessary to be able to notice and acknowledge variation, to explain it and deal with it. But if variation is indeed to be "the standard about which the statistical troops are to rally" (Wild and Pfannkuch 1999, 235), we have to arrive at a common conceptualization of statistics instruction in terms of variation. Pfannkuch (1997) discusses the characteristics of statistical reasoning laid out by a practicing as well as teaching statistician during an in-depth interview she conducted to investigate his perspective on the nature of statistical reasoning. Based on insights obtained from the interview, she offers the following epistemological triangle, which has at its core the development of an understanding of variation, as a model for introductory statistics instruction.

Figure 1. Pfannkuch's Epistemological Triangle.

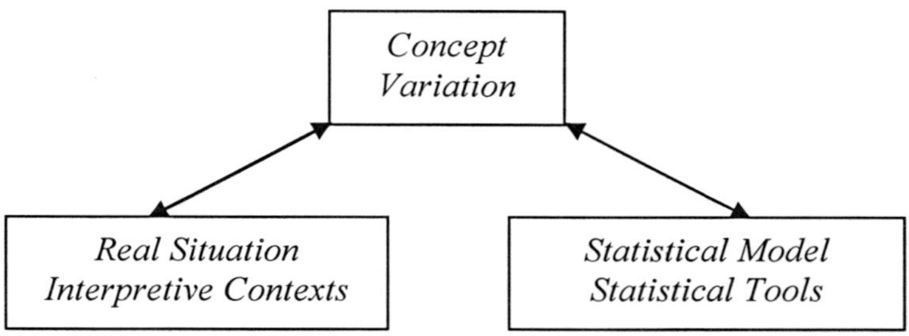

In encouraging students to develop their understanding of the concept of variation, Pfannkuch's epistemological triangle aims at the same time to promote a richer understanding of all the other main statistical ideas. The epistemological triangle indicates that for conceptualization of variation, a combination of subject and context knowledge is essential (Pfannkuch 1997). The interlinked arrows indicate the strong connection that has to be created between statistical tools and the context of the problem. The assumption underlying the epistemological triangle is that the concept of variation would be subject to development over a long period of time, through a variety of tools and contexts (Pfannkuch 1997). As pointed out earlier, emphasizing "the interplay of data and theory" is vital, since the main purpose of statistical tools such as graphs and statistical summaries is to help understand or make predictions about stochastic realworld phenomena using a statistical model of them.

Meletiou-Mavrotheris and Lee (2002) describe the experiences and insights gained from a teaching experiment that employed Pfannkuch's epistemological triangle as a guide to curriculum development and instruction. The teaching experiment, which took place in a college level introductory statistics course in the United States, implemented a nontraditional path to statistics instruction with variation at its core. The conjecture guiding the study was that the reason behind students' difficulties with the stochastic might be the instructional neglect of variation, and that if statistics instruction were to put more emphasis on helping students build sound intuitions about variation and its relevance to statistics, then we would be able to witness improved understanding of statistical concepts.

By contrasting students' intuitions about the stochastic prior to instruction with their stochastical reasoning at the completion of the course, Meletiou-Mavrotheris and Lee (2002) illustrate how the instructional approach employed in the study proved a promising alternative to more conventional instruction. Findings from student assessments at the beginning of the course further supported the conjecture that variation is neglected and its critical role in statistical reasoning is under-recognized. Students participating in the study exhibited, at the outset of instruction, a strong tendency to think deterministically and difficulties in differentiating between chance variation in the data and variation due to some form of underlying causality. This tendency was more pronounced in real world-contexts.

Assessment of student learning at the end of the teaching experiment described in Meletiou-Mavrotheris and Lee (2002) suggests that the pedagogy implemented in the experiment did promote understanding of the stochastic nature of statistical concepts and might deserve further investigation. Findings also indicate that the emphasis of instruction on the omnipresence of variation and the complementarity of theory and experience was indeed helpful in building bridges between students' intuitions, and statistical reasoning. Students were much less quick than before to assume that short-term fluctuations in the data must be causal and to develop causal explanations. The efforts of instruction to present statistical thinking as a balance between deterministic and stochastical reasoning succeeded in helping students move away from *uni-dimensional* thinking and integrate center and variation into their analyses and predictions. Although not totally letting go of their deterministic mindset, students were much more willing to interpret situations using a combination of stochastic and deterministic reasoning. The teaching experiment increased significantly students' awareness of variation and its effects. Instruction managed to get across to students the idea that "thinking about variability is the main message of statistics" (Smith 1999, 249).

5.2 Changing the Culture of the Mathematics Classroom

The research literature indicates a strong deterministic mindset among adults, which is extremely difficult to change. In order to ensure that sound intuitions about the stochastic become "part of the permanent intellectual bloodstream of the student" (Kettenring 1997, 153), reformers should concentrate efforts on finding ways to develop students' statistical reasoning at a young age. I am convinced that if mathematics instruction started offering to children from a very early age an alternative learning environment that would nourish the development not only of deterministic but also of probabilistic reasoning, people would be able to reason much more effectively about the stochastic. Although stochastics has already been established as a vital part of the K-12 mathematics curriculum in many countries, instruction of statistical concepts is, as at the college level, still highly influenced by the formalist mathematical tradition. In order to move away from that tradition, there ought to be fundamental changes to the curricular materials, instructional methods, tools, and cognitive technologies employed in the classroom to teach statistical and probabilistic concepts.

One of the most important factors in educational change is, in fact, the change in teaching practices. The extent to which students assume that deterministic causality underlies variation, as opposed to the possibility of random variation, depends in part upon the orientation of their classroom culture (Metz 1997). The culture of the classroom, in the activities it structures and the interpretations it values, might embody a deterministic or a non-deterministic view of the world (Metz 1997). Teachers' choices in their instruction are a central factor in students' learning. For it is what a teacher knows and can do that influences how she or he organizes and conducts lessons, and it is the nature of these lessons that ultimately determines what students learn and how they learn it. Teachers who currently teach statistics, however, have, for the most part, received a purely formalistic education, holding degrees in pure mathematics, with little if any training in stochastic reasoning. Thus they are likely to have relatively deterministic epistemological sets. Reformers in mathematics education are calling for efforts to retrain teachers in the new paradigm, not only through statistics coursework, but also in appropriate pedagogical courses that may help them understand the difficulties young students face with respect to stochastical reasoning.

A second important factor in educational change is the development of appropriate curricula that incorporate ideas of stochastical reasoning in activities appropriate for young students. Currently very few curricula make an effort to develop stochastical reasoning in a systematic manner through the years. For example, activities that involve exploratory data collection can

be introduced as early as pre-school. Children may collect, organize, and display data that are relevant to their interests and lives from an early age and may be encouraged to develop conjectures and hypotheses based on these data. As students progress through the grades, the sophistication of their data collection techniques, their organization and display of data, as well as the interpretations and conjectures they may reach will have to increase appropriately. Hence questions about issues such as variability, representativeness, and patterns may arise naturally over time; and students may have the opportunity to explore these issues in contexts that are both rich and meaningful to them.

6. IMPLICATIONS FOR FURTHER RESEARCH

Statistics education research could significantly contribute towards finding ways to help learners build powerful connections between formal mathematical expressions of the stochastic and everyday informal intuitions (Borovcnik 1990). Longitudinal studies that trace the evolution over time of students' fundamental statistical concepts (such as understanding of variation as a consequence of the interaction between mathematics and statistics curricula and instruction) would be extremely helpful towards discovering the sources of student difficulties with the stochastic. Such studies should investigate the relationship between stochastical and mathematical thinking, learning, and teaching, with regard to not only the cognitive dimension but also the epistemological and cultural dimensions (Metz 1997). The epistemological dimension involves looking at the effect of students' beliefs about the place of chance and uncertainty on their emergent understandings. The cultural dimension involves investigating how messages about the place of chance and determination, implicit in the practices and values of the classroom, influence students' beliefs and ideas. Indicators of the classroom culture (Metz 1997) that should be examined include choice of subject matter, structuring of problems, the teacher's reaction to students' claims about causality, the aesthetics of what constitutes a good solution or explanation, and the teacher's willingness to accept multiple strategies and viewpoints.

REFERENCES

Azcarate, P., and Cardenoso, J. M. (1994). Why Ask Why? *Research Papers from the Fourth International Conference on Teaching Statistics.* Minneapolis: The International Study Group for Research on Learning Probability and Statistics.

Ballman, K. (1997). Greater Emphasis on Variation in an Introductory Statistics Course. *Journal of Statistics Education*, 5 (2).

Biehler, R. (1994). Probabilistic thinking, statistical reasoning, and the search for causes: Do we need a probabilistic revolution after we have taught data analysis? In J. B. Garfield (Ed.) *Research Papers from the Fourth International Conference on Teaching Statistics.* Minneapolis: The International Study Group for Research on Learning Probability and Statistics.

Biehler, R. (1997). Students' Difficulties in Practicing Computer-Supported Data Analysis: Some Hypothetical Generalizations From Results of Two Exploratory Studies. In J. B. Garfield and G. Burrill (Eds.) *Research on the Role of Technology in Teaching and Learning Statistics* (176-197). Voorburg, The Netherlands: International Statistical Institute.

Borovnik, M. (1990). A Complementarity Between Intuitions and Mathematics. In J. B. Garfield (Ed.) *Research Papers from the Third International Conference on Teaching Statistics*. University of Otago, Dunedin, New Zealand.

Borovnik, M., and Peard, R. (1996). Probability. In A.J. Bishop (Ed.) *International Handbook of Mathematics Education* (239-287). Netherlands: Kluwer Academic Publishers.

Breslow, N. E. (1999). Discussion: Statistical Thinking in Practice. *International Statistical Review*, 67(3), 252-255.

Confrey, J. (1980). *Conceptual change analysis: Implications for mathematics and curriculum inquiry.* East Lansing, MI: Institute for Research on Teaching, Science-Mathematics Teaching Center, Michigan State University.

Confrey, J. (1991). Learning to listen: A student's understanding of powers of ten. In E. von Glaserfeld (Ed.) *Radical Constructivism in Mathematics Education* (111-136). Netherlands: Kluwer Academic Publishers.

Fischbein, E. (1975). *The Intuitive Sources of Probabilistic Thinking in Children.* Dordrecht, The Netherlands: Reidel.

Garfield J. and delMas, R. (1990). Exploring the stability of students' conceptions of probability. In J. B. Garfield (Ed.) *Research Papers from the Third International Conference on Teaching Statistics.* University of Otago, Dunedin, New Zealand.

Ghosh, J. K. (1997). Discussion. *International Statistical Review*, 65 (2), 154-155.

Gilligan, C. (1986). Remappping the Moral Domain: New Images of Self in Relationship In T. Heller, M. Sosna, and D. Welber (Eds.) *Reconstructing Individualism: Autonomy, Individualism, and the Self in Western Thought.* Stanford: Stanford University Press.

Glencross, M. J., and Binyavanga, K. W. (1997). The role of technology in statistics education: A view from a developing region. In J. Garfield and G. Burrill (Eds.) *Research on the Role of Technology in Teaching and Learning Statistics* (301-308). The Netherlands: ISI.

Hawkins, A. (1997a). Myth-conceptions. In J. B. Garfield and G. Burrill (Eds.) *Research on the Role of Technology in Teaching and Learning Statistics* (vii-viii). Voorburg, The Netherlands: International Statistical Institute.

Hawkins, A. (1997b). Discussion. *International Statistical Review*, 65 (2), 141-146.

Hoerl, R., Hahn, G. and Doganaksoy, N. (1997). Discussion: Let's Stop Squandering Our Most Strategic Weapon. *International Statistical Review*, 65 (2), 147-153.

Joiner, G. L. and Gaudard, M. A. (1990, December). Variation, management, and W. Edwards Deming. *Quality Progress*, 29-37.

Kahneman, D., and Tversky, A. (1973*). On the Psychology of Prediction. Psychological Review*, 80 (4), 237-251.

Kettenring, J. (1997). Discussion. *International Statistical Review*, 65 (2), 153.

Konold, C. (1989). Informal Conceptions of Probability. *Cognition and Instruction*, 6, 59-98.

Konold, C. (1995). Issues in Assessing Conceptual Understanding in Probability and Statistics. *Journal of Statistics Education*, 3 (1).

Kuhn, T. (1962). *The Structure of Scientific Revolution.* Chicago: University of Chicago Press.

Lachance, A., and Confrey, J. (1996, April). *Mapping the Journey of Students' Explorations of Decimal Notation via Ratio and Proportion.* Presented at the Annual Meeting of the American Educational Research Association, New York, NY.

Lakatos, I. (1976). *Proofs and Refutations.* Cambridge: Cambridge University Press.

Latour, B. (1987). *Science in Action.* Cambridge, MA: Harvard University Press.

Lisbon Summit (2000, March). *Conclusions of the Lisbon Summit.* http://ue.eu.int/summ.htm

Meletiou, M. (2000). *Developing students' conceptions of variation: An untapped well in statistical reasoning*. Unpublished doctoral thesis. Austin, University of Texas: 336.

Meletiou-Mavrotheris, M., and Lee, C. (2002). Teaching Students the Stochastic Nature of Statistical Concepts in an Introductory Statistics Course. *Statistics Education Research Journal*, 1 (2), 22-37.

Meletiou-Mavrotheris, M., and Lee, C. (2003). *Students' Conceptual Development of Variation: A Study Using the Transformative and Conjecture-Driven Methodology.* Presented at: Third International Research Forum on Statistical Reasoning, Thinking and Literacy, Lincoln, Nebraska.

Metz, K. E. (1997). Dimensions in the Assessment of Students' Understanding and Application of Chance. In I. Gal and J. B. Garfield (Eds.) *The Assessment Challenge in Statistics Education.* Burke, VA: IOS Press.

Moore, D. (1990). Uncertainty. In L.Steen (Ed.) *On the shoulders of giants: new approaches to numeracy* (95-137). USA: National Academy Press.

Moore, D. (1997). New Pedagogy and New Content: The Case of Statistics. *International Statistical Review*, 65 (2), 123-165.

National Council of Teachers of Mathematics (2000). *Principles and standards for school mathematics*. Reston, VA: The author.

Nickson, M. (1981). *Social foundations of the mathematics curriculum: A rationale for change.* Unpublished doctoral dissertation, Institute of Education, University of London.

Nickson, M. (1992). The culture of the mathematics classroom: An unknown quantity? In D. A. Grows (Ed.) *Handbook of Research on Mathematics Teaching and Learning* (101-114). New York: Macmillan.

Nisbett, R., Krantz, D., Jepson, C., and Kunda, Z. (1983). The Use of Statistical Heuristics in Everyday Inductive Reasoning. *Psychological Review*, 90 (4), 339-363.

Pfannkuch, M. (1997). Statistical Thinking: One Statistician's Perspective. In J. Garfield and J. Truran (Eds.) *Research Papers on Stochastics Education*, 171-178.

Pfannkuch, M. and Brown, C. M. (1996). Building on and Challenging Students' Intuitions About Probability: Can We Improve Undergraduate Learning? *Journal of Statistics Education*, 4 (1).

Philips, B. (1999, November). Report from the IASE President. *IASE Review*, 2-3.

Pólya, G. (1962). *Mathematical Discovery.* New York: John Wiley and Sons.

Pratt, D. C. (1998). *The Construction of Meanings In and For a Stochastic Domain of Abstraction.* Ph.D. Thesis, University of London.

Rubin, A., Bruce, B., and Tenney, Y. (1990). Learning about Sampling: Trouble at the Core of Statistics. In J. B. Garfield (Ed.) *Research Papers from the Third International Conference on Teaching Statistics.* University of Otago, Dunedin, New Zealand.

Scarano, G. H., and Confrey, J. (1996, April). *Results from a Three-Year Longitudinal Teaching Experiment Designed to Investigate Splitting, Ratio and Proportion.* Presented at the Annual Meeting of the American Educational Research Association, New York, NY.

Shaughnessy, J. M. (1992). Research in Probability and Statistics: Reflections and Directions. In D. Grouws (Ed.) *Handbook of Research on Mathematics Teaching and Learning* (465-494). New York: Macmillan.

Shaughnessy, J. M. (1997). Missed opportunities on the teaching and learning of data and chance. In J. Garfield and J. Truran (Eds.) *Research Papers on Stochastics Education* (129-145).

Shaughnessy, J., and Ciancetta, M. (2001). Conflict Between Students' Personal Theories and Actual Data: The Spectre of Variation. Paper presented at the Second Roundtable Conference on Research on Statistics Teaching and Learning, Armidale, New South Wales, Australia.

Shaughnessy, J. M., Watson, J., Moritz, J., and Reading, C. (1999, April). *School Mathematics Students' Acknowledgment of Statistical Variation.* Presentation at the NCTM Research Presession, San Francisco, CA.

Smith, T. M. F. (1999). Discussion. *International Statistical Review*, 67 (3), 248-250.

Snee, R. D. (1999). Discussion: Development and Use of Statistical Thinking: A New Era. *International Statistical Review*, 67 (3), 255-258.

Steinbring, H. (1990). The Use of Chance-Concept in Everyday Teaching - Aspects of a Socially Constituted Epistemology of Mathematical Knowledge. In J. B. Garfield (Ed.) *Research Papers from the Third International Conference on Teaching Statistics.* University of Otago, Dunedin, New Zealand.

Wiesman, D. and Wotring, E. (1997). *The Scientific, Humanistic, and Critical Paradigms: A Second Look* [Online]. Available: http://mailer.fsu.edu/~ewotring/com5312/ notes.html.

Wild, C. J., and Pfannkuch, M. (1999). Statistical Thinking in Empirical Enquiry. *International Statistical Review*, 67, 3, 223-265.

Wilensky, U. (1993). *Connected Mathematics - Building Concrete Relationships with Mathematical Knowledge.* Ph.D. Thesis, Massachusetts Institute of Technology.

Wilson, B., Teslow, J., and Osman-Jouchoux, R. (1995). The Impact of Constructivism (and Postmodernism) on ID Fundamentals. In B. Seels (Ed.) *Instructional Design Fundamentals: A Review and Reconsideration* (137-157). Englewood Cliffs NJ: Educational Technology Publications.

INTERLUDE 7

"Are intuitions going berserk?" was the last sentence of the previous Interlude. The next contribution by Ard Van Moer wants to address the famous problem of what intuition actually is. We need to be careful here, because *intuition* tends to appear in quite different contexts and it is a good thing to keep them, at least in an initial phase, separated. Meletiou-Mavrotheris calls it intuitions when we think *spontaneously* (by the way, what are our intuitions about that magical term?) about specific mathematical notions. There is apart from that also the intuition that professional mathematicians talk about, the famous or infamous mathematical intuition that for some, Kurt Gödel and Roger Penrose being the best known examples, constitutes a direct access to mathematical heaven, the Platonic realm of ideal forms, concepts, and ideas. That very same intuition allows mathematicians to discover (or to invent, or to construct?) their proofs.

We hear the reader protesting: "Please could you tell us something new? We know about Platonism and the like, we know about discovery versus invention in the philosophy of mathematics." Many answers are possible. How about the following idea? Suppose you agree that a particular kind of philosophical position about mathematics must have some reflection on how mathematics is taught, e.g., in secondary schools. If so, please consider the following statement: "Philosophically, I consider myself to be a [____], hence my preferred theory or view about mathematics education is [____]"; the idea being that, if the first blank is filled in, it becomes immediately clear what one should write for the second blank. We are convinced that very, very few among us have ever done this particular exercise. True, for some positions, such as formalism, constructivism, Platonism (although we have doubts about this one), the exercise has been done; but what if you feel inclined towards a sociological understanding of mathematics? This surely is less trivial.

Do note that these considerations, as important as they may are, do not answer the even more important question whether intuition can be taught or not. A major part of Van Moer's paper is dedicated precisely to that problem. Is it a matter of "Either you have it or you don't" or is it a matter of training? Yes, this question too has been posed over and over again, but did we ever get a satisfactory answer?

LOGIC AND INTUITION IN MATHEMATICS AND MATHEMATICAL EDUCATION

Ard Van Moer
Free University of Brussels, Belgium

Abstract: A good mathematics teacher is not only a good mathematician, but also a good teacher. In other words, a good mathematics teacher is not only able to solve mathematical problems, (s)he is also able to explain how mathematical problems are solved. Many mathematicans (and mathematics teachers) are, however, able to solve mathematical problems without knowing or understanding *how* they solve these problems: solving a mathematical problem often involves a multitude of unconscious or intuitive mental processes. And a person who solves certain problems without knowing or understanding how (s)he solves these problems is not able to explain to others how these problems can be solved. Consequently, many mathematics teachers would be better teachers if they knew more about the psychology of mathematicians and mathematical invention. In this article, the distinction between mathematical invention and mathematical discovery will be discussed from a psychological viewpoint. Some ideas about the psychology of mathematicians and mathematical invention will be formulated. These ideas fit in with so-called *universal Darwinism* and will be helpful in understanding the distinction between mathematical intuition on the one hand, and deduction or logic on the other.

Key words: Discovery, invention, deduction, intuition, mathematical education

1. DISCOVERY AND INVENTION IN MATHEMATICS

Cars are very important in modern society. Most people use cars regularly, even if they have no idea at all—or only a very vague idea—how or why a car works. Some people, however, know more about cars. Most car mechanics, for instance, have more than just a vague idea how or why a car

works: they are familiar with the different parts of cars, they understand how and why these different parts are interconnected, and they know a lot about the processes that take place in some of these parts and about the way some of these parts are structured or built up. And then there are some lucky persons who are actually able to *construct* cars: these people design cars, invent new ways to structure or build up certain parts of cars, and sometimes even invent completely new parts or new kinds of cars.

Mathematics is also very important in modern society. Most people use mathematics regularly, even if they have no idea at all—or only a very vague idea—how or why mathematics works. Some people, however, know more about mathematics. Most mathematicians and mathematics teachers, for instance, have more than just a vague idea how or why mathematics works: they are familiar with the different parts of mathematics, they understand how and why these different parts are interconnected, and they know a lot about the processes that take place in some of these parts and about the way some of these parts are structured or built up. And then there are some lucky persons who are actually able to *construct* mathematics: these people design mathematics, invent new ways to structure or build up certain parts of mathematics, and sometimes even invent completely new parts or new kinds of mathematics.

There is, however, an important difference between cars on the one hand, and mathematics on the other. It seems that there is, in most human beings, a natural tendency to like cars and to dislike mathematics. After all, cars are concrete objects that seem to enhance our liberty and to enable us to move freely from one place to another (when we're not stuck in a queue), whereas mathematics is often regarded as abstract stuff that limits—in one way or another—our capacity to think freely and creatively, as if it is possible that an overdose of mathematics would reduce all our thought processes to some weird or spooky kind of purely deductive reasoning that resolves itself into a servile and blind process of observing certain given rules. In this article, I will argue that this way of looking at mathematics is completely wrong.

It is, however, a fact that many people are averse to mathematics. Some people are even downright afraid of mathematics. Since it is almost impossible to impart mathematical knowledge to a student who is afraid of mathematics, one of the most important—and difficult—tasks of mathematics teachers is to try to *motivate* their students to do mathematics.

Some students do not like mathematics, but understand that mathematics is very important in modern society. These students regard mathematics as some kind of necessary evil. As a consequence, they just want to know how to *use* mathematics, because they realize that this knowledge may come in handy later on. This is, however, not good enough. We do not only need students who want to know how to *use* mathematics, we also need some

students who want to know how or why mathematics works. We even need some students who want to know how mathematics can be constructed and how new parts or new kinds of mathematics can be invented. It is, therefore, necessary to teach at least some students how inventions are effected in the mathematical field and to give them the opportunity to experience the "kick" of mathematical invention by themselves. This is another important—and difficult—task of mathematics teachers. Although some mathematics teachers neglect this task, it is, in my view, as important as their task to motivate their students to do mathematics. I also think it is even more difficult.

Before I pay closer attention to the challenges mathematics teachers are faced with, I want to elucidate my use of the word *invention* in this (mathematical) context. Some philosophers assert that mathematics is, all things considered, a human *invention.* Other philosophers, however, assert that mathematics is not invented but *discovered.* This raises the interesting question of which properties or attributes of inventions and discoveries scientific, artistic, or other really constitute the difference between an invention and a discovery.

It is often argued that a discovery of a fact, a law, a thing, or a place is always a discovery of a fact, a law, a thing, or a place that already existed before it was discovered, whereas an invention is always an invention of something new that did *not* exist before it was invented. We say that Columbus *discovered* America, because America already existed before it was discovered by Columbus. We also say that Alexander Graham Bell *invented* the telephone, because there were no telephones before Bell invented them. Inventions, unlike discoveries, bring about new kinds of things.

This difference between discoveries and inventions is, however, not always crystal-clear. Discoveries and intentions often go hand in hand. For instance, Toricelli *discovered* that, when one inverts a closed tube on a mercury trough, the mercury ascends to a certain determinate height, but in doing this he *invented* the barometer (Hadamard 1945, xi).

We also encounter serious ontological problems if we try to apply the above-mentioned distinction between discoveries and inventions to mathematical concepts, ideas, and theorems. Did Newton and Leibniz *discover* or *invent* the integral and the integral calculus? If it is true that inventions, unlike discoveries, bring about new kinds of things, this question narrows down to the ontological question whether the integral and the integral calculus already existed before Newton and Leibniz or not. It may seem obvious that there were no integrals and no integral calculus before Newton and Leibniz, and that for this reason Newton and Leibniz *invented* them. But did Newton also *invent* the binomial theorem? Did Euler *invent*

Euler's formula? Did Euler *invent* the number e? Did mathematicians *invent* the theorem that $\sqrt{2}$ is an irrational number? Did mathematicians, for that matter, *invent* the equality $2 + 2 = 4$?

Applying the above-mentioned distinction between discoveries and inventions to mathematical concepts, ideas, and theorems is, at the very least, problematic. I will argue that the question whether mathematicians discover mathematical concepts, ideas, and theorems or invent them, is not an *ontological* question, but a *psychological* question.

So we should not ask ourselves the ontological question whether or not mathematical concepts, ideas, and theorems already existed before they were discovered or invented. This question inevitably leads to perpetual philosophical discussions. We should ask ourselves the psychological question whether or not mathematical concepts, ideas, and theorems are, in some way, dependent on the personality and the creativity of the mathematicians who discovered or invented them. Most mathematicians would say that they have *discovered* a certain mathematical concept, idea, or theorem if they are under the impression that it is entirely independent of their own personality and creativity, but that they have *invented* a mathematical concept, idea, or theorem if they are under the impression that it is, in some way, dependent on their personality and creativity.

According to this second distinction between discoveries and inventions, the question of whether mathematical concepts, ideas, and theorems are discovered by mathematicians or invented by mathematicians narrows down to the question of whether or not these mathematical concepts, ideas, and theorems are dependent on the personality and the creativity of the mathematicians who discovered or invented them. Well then, what is the answer to this last question? It depends. As an example, consider once again the integral and the integral calculus. According to the second distiction between discoveries and inventions, there can be no doubt that Newton and Leibniz *invented* the integral and the integral calculus. After all, Newton and Leibniz defined or described an integral in slightly different ways. We can say, therefore, that there was a *Newton-style* integral and a *Leibniz-style* integral. The Newton-style integral differed from the Leibniz-style integral, because Newton and Leibniz both defined or described the integral in their own personal and creative way. The invention of the Newton-style integral depended on the personality and creativity of Newton, whereas the invention of the Leibniz-style integral depended on the personality and creativity of Leibniz. Nowadays, mathematicians often use the so-called *Riemann-integral.* The Riemann-integral was defined or invented by the mathematician Bernhard Riemann. The invention of the Riemann-integral was most certainly dependent on the personality and the creativity of Riemann.

It is, in fact, completely impossible to simply discover a new mathematical concept. If a mathematician concocts a new mathematical concept, (s)he has to define or to describe this new concept. It is, however, inevitable that this definition or description will reflect—at least in some degree—his or her own personality and creativity. As a consequence, all new mathematical concepts are *invented.*

There is, however, a difference between new mathematical concepts on the one hand, and mathematical theorems on the other hand[1]. According to the second distinction between discoveries and inventions, mathematical theorems are *discovered.* Consider, for instance, Euler's solution of the so-called *Basel-problem*. Euler proved that $1+\frac{1}{4}+\frac{1}{9}+\frac{1}{16}+\ldots=\frac{\pi^2}{6}$. He proved this in a very original, personal and creative way. I'm pretty sure, however, that Euler never was under the impression that *the fact itself* that $1+\frac{1}{4}+\frac{1}{9}+\frac{1}{16}+\ldots=\frac{\pi^2}{6}$ was in some way dependent on his own personality or creativity. Euler must have understood that the mathematical theorem that $1+\frac{1}{4}+\frac{1}{9}+\frac{1}{16}+\ldots=\frac{\pi^2}{6}$ was, after all, just a logical and inevitable consequence of the axioms and the transition rules of the mathematical theory he was using. So Euler *discovered* the mathematical theorem that $1+\frac{1}{4}+\frac{1}{9}+\frac{1}{16}+\ldots=\frac{\pi^2}{6}$.

Euler's *proof* of this mathematical theorem is, however, most certainly dependent on Euler's personality and creativity. It is original, brilliant, pretty intuitive and daring, and, on the whole, very *Euler-like.* So Euler *invented* his renowned proof of the mathematical theorem that $1+\frac{1}{4}+\frac{1}{9}+\frac{1}{16}+\ldots=\frac{\pi^2}{6}$.

The mathematical theorem that $1+\frac{1}{4}+\frac{1}{9}+\frac{1}{16}+\ldots=\frac{\pi^2}{6}$ can be proved in many different ways. Euler proved this theorem in one particular way that reflected his own personality and creativity. In other words, he *discovered*

[1] The philosopher Yehuda Rav seems to agree with me. In his article *Philosophical Problems of Mathematics in the Light of Evolutionary Epistemology*, he writes: "I propose to argue that: (1) the *concept* of 'prime number' is an invention; (2) the *theorem* that there are infinitely many prime numbers is a discovery. [...] should the concept of prime number be considered an invention, a purely creative step that need not have been taken, while contrariwise it appears that an examination of the factorization properties of the natural numbers leads immediately to the "discovery" that some numbers are composite and others are not, and this looks like a simple "matter of fact" " (Rav, in Restivo, Van Bendegem and Fischer 1993, 97).

the mathematical theorem that $1+\frac{1}{4}+\frac{1}{9}+\frac{1}{16}+\ldots=\frac{\pi^2}{6}$ by *inventing* a proof of this theorem.

This kind of reasoning can be applied to *all* mathematical theorems and proofs. All mathematical theorems are logical and inevitable consequences of the axioms and the transition rules of certain mathematical theories. They are *not* dependent on the personality and the creativity of the mathematician(s) who proved them. Consequently, all mathematical theorems are *discovered*, not *invented*. However, mathematical theorems can be proved in many different ways. A mathematician who proves a certain theorem, necessarily proves this theorem in one particular way that reflects his or her own personality and creativity. A proof of a mathematical theorem is, therefore, always an invention. In other words, mathematicians *discover* mathematical theorems by *inventing* certain proofs of these theorems. Indeed, mathematical discoveries and mathematical inventions often go hand in hand.

I want to conclude these remarks on the difference between discoveries and inventions by pointing out that the second—or psychological—distinction is also applicable to other, non-mathematical discoveries and inventions. I'm sure Columbus never thought that the American continent—as a geographical entity—would have been any different without his magnificent personality. The American continent was just lying there, and Columbus stumbled upon it. We say that Columbus *discovered* America, because America—as a geographical entity—is in no way dependent on the personality or creativity of Columbus. It is, on the other hand, very well possible that Alexander Graham Bell thought that there would have been no telephones without him, or that, at the very least, telephones would have been different without him. We say that Alexander Graham Bell *invented* the telephone, because telephones are—at least in some degree—dependent on the personality or creativity of Bell.

Let's sum things up. I stated that mathematics teachers are faced with two important—and difficult—tasks, viz.:

- Mathematics teachers should try to motivate their students to do mathematics, and should do whatever is possible to interest them in mathematics; so that, in the end, these students may come to *like* mathematics
- Mathematics teachers should teach some bright students how inventions are effected in the mathematical field and give these students the opportunity to experience the "kick" of mathematical invention by themselves.

I also discussed two different kinds of mathematical invention, viz.:

- The invention of new mathematical concepts and ideas
- The invention of proofs of certain mathematical theorems.

What's more, it should by now be clear *why* mathematical invention often provides a certain *kick*. If a student or a professional mathematician invents a mathematical concept, idea, or proof, (s)he will—by the very definition of the concept *invention*—be under the impression that the invented mathematical concept, idea, or proof is—at least to some degree—dependent on his or her personality or creativity. In other words, (s)he will feel that the invented mathematical concept, idea; or proof is, in some way, reflecting his or her own personality or creativity. A student who acquires the ability to invent mathematical concepts, ideas, or proofs will no longer perceive mathematics as an *impersonal* science. For this student, doing mathematics will become a *personal* experience. The personal and subjective experience of inventing a mathematical proof, idea, or concept on your own is the *kick* I wrote about.

So far, I have not addressed the question of *how* we can give certain students the opportunity to experience this *kick* of mathematical invention by themselves. It is, however, impossible to teach our pupils and students how inventions are effected in the mathematical field, if we do not understand ourselves how these inventions are effected! For this reason, I will first try to figure out how professional mathematicians invent mathematical concepts, ideas, and proofs. In the second section of this article, I will give some—tentative—answers to the question of how inventions are effected in the mathematical field. In the third section, I will finally address the question of how we can give certain students the opportunity to experience the *kick* of mathematical invention.

2. LOGIC AND INTUITION IN MATHEMATICS

The French poet Paul Valéry (1871-1945) wrote: "It takes two to invent anything. The one makes up combinations; the other one chooses, recognizes what he wishes and what is important to him in the mass of the things which the former has imparted to him" (Hadamard 1945, 30; Dennett, 1978, 293).

Of course, Valéry is not saying that it takes two *persons* to invent anything. He is, in fact, saying that two different kinds of *mental processes* are needed in order to invent anything. The French mathematician Jacques Hadamard agrees with Valéry. Hadamard elucidates his views as follows:

> Indeed, it is obvious that invention or discovery, be it in mathematics or anywhere else, takes place by combining ideas. Now, there is an extremely great number of such combinations, most of which are devoid of interest, while, on the contrary, very few of them can be fruitful. [...]
> However, to find these [fruitful combinations], it has been necessary to construct the very numerous possible combinations, among which the useful ones are to be found.
> It cannot be avoided that this first operation take place, to a certain extent, at random, so that the role of chance is hardly doubtful in this first step of the mental process. [...]
> It is obvious that this first process, this building up of numerous combinations, is only the beginning of creation, even, as we should say, preliminary to it. [...][T]o create consists precisely in not making useless combinations and in examining only those which are useful and which are only a small minority. Invention is discernment, choice. (Hadamard 1945, 29-30)

Dennett relates this mental procedure, described by Paul Valéry and Jacques Hadamard, to the so-called *generate-and-test* procedure in artificial intelligence. He writes: "A ubiquitous strategy in AI programming is known as *generate-and-test*, and [the] quotation of Paul Valéry perfectly describes it. The problem solver (or inventor) is broken down at some point or points into a generator and a tester. The generator spews up candidates for solutions or elements of solutions to the problems, and the tester accepts or rejects them on the basis of stored criteria" (Dennett 1978, 81).

Valéry's description of the invention of poetry, Hadamard's conception of problem solving in mathematics, and the generate-and-test procedure in artificial intelligence are, in fact, all examples of so-called *universal Darwinism*. According to universal Darwinism, all phenomena with a certain degree of complexity are the result of a random process on the one hand and a selection process on the other hand, just as, according to Darwin, all biological evolution is the result of a number of random genetic mutations on the one hand, and natural selection on the other hand.

It is possible to apply universal Darwinism to the invention of mathematical concepts, ideas, and proofs. According to Hadamard, an invention in the mathematical field is always the result of two different kinds of mental processes, viz.:

- More or less uncontrolled and chaotic processes in which mathematical concepts are generated at random

- More or less controlled and orderly processes in which the randomly generated mathematical concepts are examined for their usefulness to solve certain mathematical problems

In the first kind of mental process, mathematical concepts are *generated.* In the second kind of mental process, mathematical concepts are *tested.* From this, it follows that every invention in the mathematical field is the result of a generate-and-test procedure. I will now discuss these two kinds of mental processes in more detail.

First consider the so-called *generator* or the mental process in which mathematical concepts are generated. Suppose that a certain mathematician is trying to prove a given theorem. When this mathematician starts thinking about the problem at hand, a multitude of mathematical concepts and ideas is generated in his or her mind. These concepts can be generated in two ways. Most of them are simply *remembered* by the mathematician. These concepts are ones the mathematician has encountered on previous occasions when (s)he was studying similar problems. However, not all ideas that are generated in the mathematician's mind are remembered. Some ideas are generated by *combining* other ideas. Hadamard claimed that invention or discovery, be it in mathematics or anywhere else, takes place by combining ideas." It is, for instance, possible that the idea of a hypercube is generated in the mathematician's mind by combining the idea of a cube and the idea of a four-dimensional space.

Now consider the so-called *tester,* the mental process in which mathematical concepts are tested. Suppose, once again, that a certain mathematician is trying to prove a given theorem. The concepts which are being generated in the mind of this mathematician when (s)he starts thinking about the problem are examined by the *tester* for their usefulness to prove the given theorem. A professional mathematician is familiar with many heuristic methods, rules, and tricks that enable him or her to distinguish useful mathematical concepts from useless ones. Of course, mathematical concepts that are very useful in one mathematical context can be completely useless in another mathematical context. Suppose, however, that, after having worked on the problem for some time, the mathematician thinks (s)he has found a proof of the given theorem by using different kinds of heuristic methods. How does the *tester* in the mathematician's mind decide whether this proof is useful or not? The answer to this question is very simple. Most proofs are only useful if they are *correct.* A useful proof is, in other words, a proof that is consistent with the axioms and the transition rules of a certain mathematical theory. So it would seem that the *tester* in the mind of a mathematician consists of different levels, and that it operates in different ways at different levels. At most levels, the *tester* examines the usefulness of

generated mathematical concepts by using different kinds of heuristic methods. At the last level, however, the *tester* examines the usefulness of a proposed proof by checking whether the proposed proof is correct or not. There is only one criterion at this last level of the selection process in the mind of a professional mathematician who is trying to prove a given theorem—the consistency of the proposed proof with the axioms and the transition rules of a certain mathematical theory.

These remarks on the generate-and-test procedure in the mind of a mathematician who is trying to prove a given theorem deepen our understanding of the distinction between mathematical discoveries and mathematical inventions. In the first section of this article, I stated that the question of whether mathematicians discover mathematical concepts, ideas, and theorems or invent them is, in fact, not an ontological question, but a *psychological* question. Most mathematicians would say that they have *discovered* a certain mathematical concept, idea or theorem if they are under the impression that it is entirely independent of their own personality and creativity, and that they have *invented* a mathematical concept, idea, or theorem if they are under the impression that it is, in some way, dependent on his or her personality and creativity. I also claimed that mathematical discoveries and mathematical inventions often go hand in hand, and that mathematicians discover mathematical theorems by inventing certain proofs of these theorems.

It's now easy to understand *why* mathematical discoveries and mathematical inventions often go hand in hand. In the mind of a mathematician who is trying to prove a given theorem, some mental processes will be *independent* of the personality and the creativity of this mathematician, whereas other mental processes will be *dependent* on the personality and the creativity of this mathematician.

First consider, once again, the *generator* or the mental processes in which mathematical concepts are generated. We have seen that the generated concepts in the mind of a mathematician who is trying to prove a given theorem are either remembered by the mathematician or generated by combining other ideas. Simply *remembering* a certain mathematical concept is not exactly a very creative mental process. A mathematician can only remember concepts (s)he has encountered on previous occasions. The ability of a mathematician to remember interesting mathematical concepts is, therefore, dependent on the mathematical knowledge of this mathematician. So it could be argued that the ability of a mathematician to remember mathematical concepts is—at least in some degree—dependent on his or her personality. Generating a new mathematical concept or idea by *combining* other mathematical concepts is most certainly a creative mental process. The ability of a mathematician to generate a new mathematical concept by

combining other mathematical concepts is, therefore, dependent on his or her personality and creativity.

Now consider, once again, the *tester* or the mental process in which mathematical concepts are tested. We have seen that the *tester* in the mind of a mathematician operates at different levels. At most levels, the *tester* examines the usefulness of generated mathematical concepts by using different kinds of heuristic methods. Most heuristic methods used by mathematicians are ones these mathematicians already used on previous occasions when they were studying similar problems. In other words, most heuristic methods used by a mathematician are heuristic methods this mathematician simply remembers. Simply *remembering* a certain heuristic method is not exactly a very creative mental process. However, it also happens that mathematicians—or, for that matter, students—find and develop their own heuristic methods. Finding or developing a new heuristic method is most certainly a very creative mental process. The ability of a mathematician to find or develop new heuristic methods is, therefore, dependent on his or her personality and creativity. Heuristic methods are used by mathematicians at most levels of the selection process. The last level of the selection process is, however, an exception. At this level, the usefulness of a proposed proof is examined by checking whether the proposed proof is correct or not. A correct proof is a proof that is consistent with the axioms and the transition rules of a certain mathematical theory. Checking whether a proposed proof is consistent with the axioms and the transition rules of a certain mathematical theory or not, is in fact a dull and almost *mindless* activity. This activity can be performed by any person who knows and understands the axioms and the transition rules of the mathematical theory. The ability of a mathematician to check whether a proposed proof is consistent with the axioms and the transition rules of a certain mathematical theory or not is, therefore, independent of his or her personality and creativity.

According to universal Darwinism, it is impossible to find a proof of a given mathematical theorem without some kind of generate-and-test procedure. Most mental processes in which mathematical concepts are generated seem to be—at least in some degree—*dependent* on the personality or the creativity of the mathematician who is trying to prove the given theorem. At the last level of the selection process, however, mathematical ideas are tested in a way that is necessarily *independent* of the personality and the creativity of the mathematician who is trying to prove the given theorem. Proving a given mathematical theorem is, therefore, always partly an invention and partly a discovery.

I will now pay closer attention to the mental processes in which mathematical concepts are *tested*. Some of these processes occur at an

unconscious—or subconscious—level. I will use the word *intuition* to refer to mental processes in which mathematical concepts are tested at an unconscious—or subconscious—level. It also happens, however, that mathematical concepts are tested at a conscious level. I will use the word *deduction* to refer to mental processes in which mathematical concepts are tested at a conscious level.

I am using the word *deduction* here for lack of a better word. It seems the word *intuition* has no obvious opposite. However, my use of the word *deduction* should not be confused with a more common use of this word as a denotation of a process of reasoning in which a conclusion follows necessarily—or logically—from the stated premises. An inference from the general to the specific is an example of such a process of reasoning.

This common use of the word *deduction* might, at first sight, seem entirely unconnected to my use of the word as a denotation of mental processes in which mathematical concepts are tested at a conscious level. It is nevertheless possible to relate the distinction between conscious mathematical thought processes and unconscious mathematical thought processes to the distinction between deductive—or logical—mathematical thought processes and intuitive mathematical thought processes. At unconscious levels of the mind, thought processes and ideas tend to be rather vague or indeterminate. Mathematicians—and other people—often use *quick* and *dirty* heuristics at the unconscious levels of their minds. These heuristics make it possible to reach conclusions very quickly. However, some of the conclusions that are reached by the use of *quick* and *dirty* heuristics may be completely wrong. In other words, processes of reasoning at unconscious levels of the mind are often flawed or illogical. Most mathematical thought processes in which conclusions follow necessarily or logically from the stated premises occur at a conscious level of the mind.

My use of the word *intuition* more or less conforms to the common use of this word. A mathematician who says that (s)he has used his or her *mathematical intuition* is, in fact, just saying that (s)he has accepted or rejected a mathematical idea without really knowing or understanding *why*. In other words, a mathematician who says that (s)he has used his or her *mathematical intuition* is aware of the fact that (s)he has accepted or rejected a mathematical idea, but is unaware of the mental processes involved. So these mental processes must have occurred at an unconscious—or subconscious—level.

Now suppose that John is trying to prove a given theorem about triangles. When he starts thinking about the problem at hand, a multitude of ideas is generated in his mind. Some of these ideas will be entirely non-mathematical. It is, for instance, possible that some thoughts concerning the beauty of his wife are generated at an unconscious level of his mind.

However, if John is really serious about trying to prove the given theorem and is concentrating on the problem at hand, these thoughts will be rejected as non-mathematical and irrelevant to any possible proof of the given theorem at an *unconscious* level of his mind. In this case, John will not be aware of the fact that some thoughts concerning the beauty of his wife were ever generated at all. Some of the ideas generated in John's mind will be mathematical, but unrelated to triangles and geometry. It is, for instance, possible that some ideas about matrices or Taylor series are generated at a certain—conscious or unconscious—level of John's mind. If John is an experienced mathematician, these ideas will be rejected at an *unconscious* level of his mind as irrelevant to any possible proof of the given theorem about triangles. In other words, John's ideas about matrices or Taylor series will, in this case, be rejected *intuitively* as irrelevant to any possible proof of the given theorem about triangles. However, if John is a rather inexperienced student, he may have to consider carefully the possibility of a connection between triangles on the one hand and matrices or Taylor series on the other hand before he will be convinced that there is no such connection. In this case, John's ideas about matrices or Taylor series will be rejected at a *conscious* level of John's mind as irrelevant to any possible proof of the given theorem about triangles.

Some of the ideas generated in John's mind will be more promising. When John hits upon a promising idea, he may have to go into details. Looking into details, however, is usually a conscious process. It is, for instance, possible that John surmises—for some reason—that the given theorem about triangles might be solved by calculating the coordinates of the point of intersection of two straight lines. Such a calculation is, under normal conditions, a process that occurs at a conscious level of the mind of a mathematician.

Suppose that John, after having worked on the problem at hand for some time, thinks he has found a proof of the given theorem. At this point, he should check whether the proposed proof is correct or not. In other words, John should examine now whether the proposed proof is consistent with the axioms and the transition rules of a certain mathematical theory or not. We have seen that this examination takes place at the last level of the *tester*. Examining whether a proposed proof is consistent with the axioms and the transition rules of a certain mathematical theory or not is, however, a rather dull process that often requires one to check lots and lots of details. Under normal conditions, this examination is, therefore, a process that occurs at a conscious level of the mind.

These conclusions about the processes in John's mind are applicable to the processes in the mind of any mathematician who is trying to solve a mathematical problem. Proving a mathematical theorem or solving a

mathematical problem always involves mental processes in which mathematical concepts are tested at an *unconscious* level as well as mental processes in which mathematical concepts are tested at a *conscious* level. In other words, proving a mathematical theorem or solving a mathematical problem always involves both *intuition* and *deduction.* I have to admit, however, that there are conditions in which the difference between an unconscious level of the mind and a conscious level of the mind can be rather obscure. In fact, no one really knows what consciousness is or how it works. Consequently, there are conditions in which the difference between *intuition* and *deduction* is not very clear.

It is very important for mathematics education that mathematicians, philosophers of mathematics, and teachers of mathematics admit the fact that mathematics is *not* a purely deductive science and that it is impossible to solve a mathematical problem without at least some mathematical intuition. Nevertheless, mathematics has often been regarded and/or presented as a purely deductive science by mathematicians, philosophers, and mathematics teachers. The allegation that mathematics is a purely deductive science can be interpreted in two different ways, corresponding to the two above-mentioned definitions of the word *deduction.*

- The allegation that mathematics is a purely deductive science can be interpreted as the statement that all mathematical thought processes occur at a conscious level of the mind of a mathematician
- The allegation that mathematics is a purely deductive science can be interpreted as the statement that all mathematical processes of reasoning are processes of reasoning in which conclusions follow necessarily or logically from the stated premises

I am convinced that both statements are false. It cannot be doubted that mathematics is *not* a purely deductive science after all.

3. MATHEMATICAL INTUITION AND MATHEMATICAL EDUCATION

There are three kinds of evidence for the fact that mathematics is not a purely deductive science.

First, and most important of all, there is psychological evidence. Consider, for instance, the story of Henri Poincaré. In a famous lecture at the *Société de Psychologie* in Paris, Poincaré described his discovery of certain important mathematical theorems:

> I wanted to represent these [Fuchsian] functions by the quotient of two series; this idea was perfectly conscious and deliberate; the analogy with elliptic functions guided me. I asked myself what properties these series must have if they existed, and succeeded without difficulty in forming the series I have called thetafuchsian.
>
> Just at this time, I left Caen, where I was living, to go on a geological excursion under the auspices of the School of Mines. The incidents of the travel made me forget my mathematical work. Having reached Coutances, we entered some omnibus to go some place or other. At the moment when I put my foot on the step, the idea came, without anything in my former thoughts seeming to have paved the way for it, that the transformations I had used to define the Fuchsian functions were identical with those of non-Euclidean geometry. I did not verify the idea; I should not have had time, as, upon taking my seat in the omnibus, I went on with a conversation already commenced, but I felt a perfect certainty. On my return to Caen, for conscience' sake, I verified the result at my leisure.
>
> Then I turned my attention to the study of some arithmetical questions apparently without much success and without a suspicion of any connection with my preceding researches. Disgusted with my failure, I went to spend a few days at the seaside and thought of something else. One morning, walking on the bluff, the idea came to me, with just the same characteristics of brevity, suddenness and immediate certainty, that the arithmetic transformations of indefinite ternary quadratic forms were identical with those of non-Euclidean geometry. (Hadamard 1945, 13-14)

It would seem that a great deal of Poincaré's mathematical thought processes occurred at unconscious levels of his mind. Consequently, not all mathematical thought processes occur at a conscious level of the mind of a mathematician.

There is also evidence for the fact that not all mathematical processes of reasoning are ones in which conclusions follow necessarily from the stated premises. Consider, for instance, the famous Indian mathematician Srinivasa Ramanujan. Ramanujan discovered many beautiful and amazing mathematical theorems and formulae in a rather mysterious and intuitive way. It would seem that the theorems and formulae discovered by Ramanujan were not the result of mathematical processes of reasoning in which conclusions followed necessarily from the stated premises. As a matter of fact, Ramanujan often had no idea at all how the theorems and formulae he discovered could be proved correctly.

Many mathematicians and philosophers agree that mathematics is not a purely deductive science and that it would be a mistake to deny the importance of intuition in the discovery and the invention of mathematics. Jacques Hadamard writes:

> This carries [...] the consequence that, strictly speaking, there is hardly any completely logical discovery. Some intervention of *intuition* issuing from the unconsciousness is necessary at least to initiate the *logical* work. (Hadamard 1945, 112; my emphasis)

According to Poincaré, there are two sorts of mathematicians:

> The one sort are above all preoccupied by *logic*; to read their works, one is tempted to believe they have advanced only step by step, after the manner of a Vauban who pushes on his trenches against the place besieged, leaving nothing to chance. The other sort are guided by *intuition* and at the first stroke, make quick but sometimes precarious conquests, like bold cavalrymen of the advance guard. (Poincaré, in Hadamard 1945, 106; my emphasis)

And Morris Kline writes:

> What then is mathematics if it is not a unique, rigorous, logical structure? It is a series of great *intuitions* carefully sifted, refined, and organized by the *logic* men are willing and able to apply at any time. [...]
>
> Several of the schools have tried to enclose mathematics within the confines of man's logic. But *intuition* defies encapsulation in *logic*. The concept of a safe, indubitable, and infallible body of mathematics built upon a sound foundation stems of course from the dream of the classical Greeks, embodied in the work of Euclid. This ideal guided the thinking of mathematicians for more than twenty centuries. But apparently mathematicians were misled by the "evil genius" Euclid. (Kline 1980, 312; my emphasis)

Hadamard, Poincaré, and Kline all seem to think that mathematics is more than just logic and that it is only possible to do mathematics if logic, or deduction, is supplemented with some kind of mathematical intuition.

There is also historical evidence for the fact that mathematics is not a purely deductive science. If mathematics really *were* a purely deductive science, and all mathematical processes of reasoning really *were* processes in which conclusions follow necessarily from the stated premises, then it would be impossible to reach a wrong conclusion or to deduce a false theorem from true premises. The history of mathematics is, however, replete with false theorems, false proofs, and false conjectures. Fermat's conjecture that every

number $2^{\left(2^n\right)}+1$ (with n a natural number) is a prime number, is an example of a false conjecture. The theorem that a function $f(x)$ is derivable in x_0 if it is continuous in x_0, was *proved* in many mathematical handbooks of the nineteenth century. This theorem is, however, a false theorem. The *proofs* of this theorem in the handbooks were, of course, false proofs.

What's more, there have been many paradoxes and contradictions in the history of mathematics. The ancient Greeks thought, for some reason, that irrational numbers are paradoxical. And Russell's *paradox* is a notorious example of a contradiction. This contradiction in the naive set theory of Cantor and Frege was discovered by Bertrand Russell.

There is still another reason why mathematics cannot be a purely deductive science. If it were, mathematicians who want to construct and to develop a new mathematical theory would start by defining or describing certain basic concepts of the theory and by formulating certain axioms and transition rules. This would enable them to define other more complex concepts and to prove many interesting and uninteresting theorems. In reality, however, things often seem to go the other way around. It sometimes happens that mathematicians are constructing or developing a new mathematical theory without even realizing it themselves. The construction or development of a mathematical theory often starts with a mathematical problem or the formulation of certain mathematical conjectures. The words, symbols, or concepts used to formulate these conjectures are often ill-defined. A mathematician who wants to prove the formulated conjectures needs to define these words, symbols, or concepts by reducing them to other more terms. He or she also needs to determine the transition rules of the new mathematical theory. In other words, (s)he also needs to determine the rules that may be used to prove theorems. The formulation of the axioms is often one of the last steps in the development of a mathematical theory!

Mathematics is, after all, an instance of problem solving. Mathematical problems, *not* sets of axioms and sets of transition rules, are the real drive of mathematics. If mathematics really *were* a purely deductive science, it would always go *forward*, from the basic concepts, the axioms, and the transition rules of a mathematical theory, via other more complex concepts and proofs of mathematical theorems, to the solution of all sorts of mathematical problems. In reality, however, mathematics often goes "backward", from a given mathematical problem, via the concepts and the theorems that are needed in order to solve this problem, to the basic concepts, the axioms, and the transition rules of a new mathematical theory.

So mathematics is not a purely deductive science. This raises an interesting question. If mathematics is not a purely deductive science, why is it so often *regarded* and *presented* as a purely deductive science by

mathematicians, philosophers, and mathematics teachers? There are undoubtedly many different reasons for this.

First of all, people who believe that mathematics is a purely deductive science also tend to believe that mathematics is an exact science, and that all conclusions reached by the use of mathematics are absolutely certain. It sometimes happens that other scientists are envious of mathematicians because of the supposed exactness and absolute certainty of mathematics. This exactness and absolute certainty are, however, little more than a myth.

In the first section of this article, I stated that all mathematical theorems are inevitable consequences of the axioms and the transition rules of certain mathematical theories, and that these theorems are independent of the personality and the creativity of the mathematician(s) who proved them. It is possible that a mathematician who has proved a certain theorem is so impressed with the inevitability of the theorem that (s)he forgets his or her own contribution to the proof and is apt to think that not only the theorem itself, but also its proof is independent of his or her personality and creativity. It is in other words possible that a mathematician who has discovered a certain theorem neglects the fact that (s)he has invented the proof of this theorem. A mathematician who remembers that (s)he has discovered a certain theorem but has forgotten how (s)he has invented the proof of this theorem, will probably realize that this theorem is a logical and inevitable consequence of the axioms and the transition rules of the mathematical theory (s)he was using; although (s)he may not remember that (s)he proved this theorem by the use of his or her mathematical intuition and all sorts of *quick* and *dirty* heuristics at unconscious levels of his or her mind.

In the second section of this article, I defined or described the words *deduction* and *intuition. Deduction* was described as "the mental processes in which mathematical concepts are tested at a conscious level". Intuition was described as "the mental processes in which mathematical concepts are tested at an unconscious level." So mathematicians are only aware of their deductive mathematical thought processes. They are unaware of their intuitive mathematical thought processes. It is, therefore, only logical that many mathematicians regard mathematics as a purely deductive science[2].

[2] It is still very difficult to study unconscious mental activity scientifically. In research on mathematcal thinking, researchers often use the *talking aloud procedure* and/or the so-called *clinical interview technique*. Both methods focus on conscious mental activities. Although some behavioral observations may be made in the *talking aloud procedure*, the data gathered from this method are mainly the verbalizations of a subject who is instructed to say anything that comes to his or her mind while solving a mathematical problem. And the *clinical interview technique* is, nor surprisingly, an interview technique; all data gathered from this method are entirely verbal. These verbalizations represent only what

There are, however, some mathematicians who realize that they often accept or reject certain mathematical ideas without knowing *why* they accept or reject them. Most of these mathematicians also realize that they accept and reject mathematical ideas at an unconscious level of their mind. So mathematicians are beginning to realize that the unconscious plays a part in mathematics. The unconscious is, as a matter of fact, a rather *new* discovery.

It is easier to *teach* mathematics as a deductive science. There are two reasons for this. First, it is easier to teach students to go *forward*, from given axioms, transition rules, and basic concepts, via more complex concepts and proofs of mathematical theorems, to the solution of mathematical problems, than to teach them to go *backward* from given mathematical problems, via the complex concepts and the theorems that are needed in order to solve these problems, to the comparatively simple basic concepts, axioms, and transition rules. Second, it is easier to teach students some clear-cut concepts, axioms, and rules they can use consciously than to teach them how they can develop and use their mathematical intuition and all sorts of *quick* and *dirty* heuristics at unconscious levels of their minds.

It seems we have come full circle. This article began with some general remarks on the education of mathematics. Now, at the end of the article, I will try to explain why it is so important for the education of mathematics that mathematicians, philosophers of mathematics, and teachers of mathematics admit the fact that mathematics is not a purely deductive science. I will also try to answer the question of how we can give certain pupils and students the opportunity to experience the *kick* of mathematical invention by themselves.

I think it is a very bad idea to teach mathematics as a purely deductive science, if it is, in fact, not a purely deductive science at all. Pretending that mathematics is a purely deductive science constitutes a lie. I know from experience that it is very tempting for mathematics teachers to present mathematics as an exact science that offers absolute certainty. Some mathematics teachers almost pose as deities by their benevolence of offering their students a glimpse of the absolute and eternal truth and certainty of mathematics. However, students have a right to know that the history of mathematics is replete with false theorems, false proofs, false conjectures, paradoxes, and contradictions.

There is another, equally important, reason why mathematics should not be taught as a purely deductive science. In the first section of this article, I stated that most people can function normally in daily life if they just know how to *use* certain mathematical concepts. Nevertheless, we also need some

goes on at the conscious levels of the mind of the subject. (See the article of Ginsburg, Kossan, Schwartz, and Swanson in Ginsburg 1983, 7-16.)

people who know how or why mathematics works, how mathematics can be constructed, and how new parts or new kinds of mathematics can be invented. For this reason, mathematics teachers should give bright pupils and students the opportunity to experience the *kick* of mathematical invention by themselves. A student who invents a mathematical concept, idea or proof will often experience some kind of *kick*, because a person who *invents* a mathematical concept, idea, or proof is under the impression that it at least to some degree, dependent on his or her personality or creativity.

In the second section of this article, I stated that the ability of a mathematician to check whether ore not a proposed proof is consistent with the axioms and the transition rules of a certain mathematical theory is independent of the personality and creativity of the mathematician. The ability of a mathematician to find and develop new heuristic methods is, however, most certainly dependent on his or her personality and creativity. If mathematics is taught as a purely deductive science and pupils and students are told only how they should perform certain dull and *mindless* activities, then the students will perceive mathematics as an impersonal science. In this case, students are not given the opportunity to experience the *kick* of mathematical invention by themselves. If, however, mathematics is *not* taught as a purely deductive science and students are also helped to develop their own heuristics and mathematical intuition, then students will have a chance to experience the *kick* of mathematical invention by themselves.

For this reason, mathematics teachers should help their students to develop their mathematical intuition. This is not an easy task. I have already stated that it is easier to teach students some clear-cut concepts, axioms, and rules they can use consciously than to teach them how they can develop and use their mathematical intuition and all sorts of *quick* and *dirty* heuristics at unconscious levels of their minds. The task is, however, not an impossible one. Mathematics teachers can help their students to develop their mathematical intuition by the use of certain techniques. For instance, the list of questions published by George Pólya in his book *How to Solve It* (1945) can be used by mathematics teachers to help their students solve mathematical problems. Mathematics teachers can help their students develop their mathematical intuition by posing them these questions. Many philosophers, including Imre Lakatos, were influenced by Pólya. These philosophers understood the importance of heuristics and of the so-called "context of discovery" (Van Kerkhove. See this volume). However, mathematical problem solving in the framework of mathematics education is still of current interest. Much work remains to be done in this area.

REFERENCES

Benacerraf, Paul and Putnam. (1964). *Philosophy of mathematics: Selected readings*. Cambridge University Press, Reprinted in 1985.

Dennett, Daniel C. (1978). *Brainstorms: Philosophical Essays on Mind and Psychology*. Cambridge, Mass.: MIT Press/A Bradford Book.

Dennett, Daniel C. (1995). *Darwin's Dangerous Idea:Evolution and the Meanings of Life*. New York: Simon and Schuster.

Ginsburg, Herbert P. (Ed.) (1983). *The Development of Mathematical Thinking*. London: Academic Press.

Hadamard, Jacques. (1945). *The Psychology of Invention in the Mathematical Field*. Princeton University Press, Reprinted in 1954.

Kanigel, Robert. (1991). *The Man Who Knew Infinity. A Life of the genius Ramanujan*. Washington Square Press.

Kline, Morris. (1980). *Mathematics: The Loss of Certainty*. New York: Oxford Univerity Press

Pólya, George. (1945). *How to Solve It: A New Aspect of Mathematical Method*. Princeton University Press.

Restivo, Sal, Jean Paul Van Bendegem and Roland Fischer (Eds.) (1993). *Math Worlds. Philosophical and Social Studies of Mathematics and Mathematics Education*. New York: State University of New York Press.

Royer, James M. (ed.) (2003). *Mathematical Cognition*. Greenwich, Connecticut: Information Age Publishing.

INTERLUDE 8

We human beings are particularly fond of making distinctions. If we are to believe Spencer-Brown in his little known *Laws of Form*, that is the beginning of all our thinking about and acting in the world. No doubt it is a good thing to distinguish the wall from the door in the wall, to see the difference between high and low when jumping out of the window. That does not, however, prevent us from making unnecessary, unhelpful, or worse still, counterproductive distinctions. To a large extent, what Van Moer has done in his paper is, almost unnoticed, to ignore such a counterproductive distinction, one that has managed to become one of the most basic boundary lines in the present-day philosophy of science and, by extension, in the present-day philosophy of mathematics, viz., the distinction between the *context of discovery* and the *context of justification*. Something all of us have known for years.

That being said, it is our feeling that one aspect of this distinction has been and still is usually ignored: it is generally accepted that by making this distinction and by focusing on only one of the two distinctive things, in almost all cases the context of justification, one is driven towards specific philosophies of science. Take the most famous example: focusing on justification allows the use of logical methods as ingredients for a potential methodology of science, for does not the context of discovery require the presence of irrational elements that logic shuns away from? This we all know, but what about the reverse situation: if you are willing to erase the distinction, if you are willing to treat both contexts as necessary elements of a full-fledged theory of what it is mathematicians do when they claim that they are doing mathematics, what will be the effect on your philosophical view(s)?

Bart Van Kerkhove performs such an exercise in his contribution. Taking Imre Lakatos as a starting-point, taking his quasi-empiricism seriously and thinking it through, relying on, among others, Paul Ernest, Eduard Glas, and Teun Koetsier, it soon becomes apparent that it makes sense to talk about a humanistic mathematics. Presenting mathematics to school children as a human enterprise and not as a godly affair does indeed have profound repercussions on one's view of what mathematics is. So, at least in some cases, the philosopher has a lesson to learn.

A PLACE FOR EDUCATION IN THE CONTEMPORARY PHILOSOPHY OF MATHEMATICS

The Case of Quasi-Empiricism

Bart Van Kerkhove
Free University of Brussels – Center for Logic and Philosophy of Science, Belgium

Abstract: The paper examines the possibility of a positive role for the educational perspective in progressive philosophy of mathematics. In particular, dialectalism in mathematics, as initiated by Imre Lakatos under the inspiration, among others, of the giant mathematics educator George Pólya, is looked into. Important "humanistic" issues such as fallibility and plausibility are addressed. Although still requiring further scrutiny, the quasi-empiricism presented at the very least proves to be a viable contender to the various mainstream *a priori* philosophical positions concerning the nature of mathematical knowledge/science.

Key words: Dialectalism, quasi-empiricism, fallibility, plausible thinking, Pólya, Lakatos

1. INTRODUCTION

In this paper, we shall trace a particular connection between mathematics education and mathematical humanism, a philosophical school broadly concerned with mathematics as a distinctively human cultural product, and therefore with the practice of mathematicians as epistemologically relevant. As an entry into this subject, we consider the statement of purpose of the Humanistic Mathematics Network Journal (HMNJ). Founded in 1987 as a newsletter following a conference, HMNJ evolved as an irregularly published periodical, which in 2002 went exclusively online. It was started with the intention, reprinted at the beginning of every issue, of reflecting on and promoting what were thought of as two intimately related themes: the

development of humanistic mathematics, and the humanistic teaching of mathematics. There is no question that the latter cannot exist without the former. But what about the other way round? This will be our topic of interest. That is, in what ways, if any, does, can, or should pedagogical settings contribute to the (re)development of mathematics and its philosophical (re)assessment? In particular, we shall focus on how the pedagogical perspective can teach us about the extent to which mathematical progress is of a dialectical nature, and how this can accordingly be of service in building proper humanistic reconstructions of mathematics, which "relate mathematical discoveries to personal courage, discovery to verification, mathematics to science, truth to utility, and in general, mathematics to the culture within which it is embedded" (from the HMNJ manifesto).

Note that our chosen option does not involve an exclusivity claim: humanistic mathematics might indeed be interpreted differently, and thus be associated with a number of different alternative schools in the philosophy of mathematics. In various ways, these are all humanistic alternatives to absolutism (regularly combined with an all-pervasive Platonism) about mathematics, which says: there is conceivably something like total certainty in mathematics, and it is the philosopher's job to look for and secure it. Humanism, on the contrary, brings in an element of fallibility, the philosopher's task in this case being to challenge absolutism and develop another type of foundational thinking, viz., one more broadly conceived, and questioning what is at the basis of mathematical knowledge, as being *un*limited to formal or logical foundations. Constructivists, for example, would stress the fact that mathematics is a product of the human mind. Intuitionists, in the wake of Brouwer, even envisaged that it be built up from scratch on an absolutely *a priori* or empirically neutral mental basis—a radical view which has been widely disapproved of for being utterly unrealistic, viz., dismissing the bulk of spectacular mathematical results (e.g., rejecting the infinite). Less radical spirits will therefore at the very least postpone this dismissal until it has been conclusively shown that the main results of modern mathematics are roughly translatable in their proposed constructivist systems. *Humanism* might also be taken in another sense, in which it means "the service of mankind and its set goals," mathematics thus above all having to be applicable, practical, instrumental, pragmatical, convenient. One would then associate it not with a constructivist, but rather with a conventionalist picture, in the style of Hilbert or the logical positivists: pick your axiom system, in accordance with what the mathematics should be capable of. A similar stance yields a philosophically rare combination of *a posteriori* and *a priori*, for actual problems determine the choice of system, while after this choice, one continues within the system without any external interference.

A third option, to be traced here, removes the *a priori* component altogether: empiricists consider mathematics to result essentially from human sense experience, viz., as the most extreme form of generalization and abstraction. As an important prior remark, a disclaimer of reproaches of antifoundationalism, please note that empiricism does not require radical reform either, although it does create room, at least in principle, for possible alternatives to prevalent *universal* mathematics. It seems especially to give rise to thought experiments in the style of: Had our senses evolved somewhat differently, would mathematics have been anything like what it is now? Although philosophically important, possibilities like these remain hypothetical and are being formulated above all to underscore the humanism involved. Contrary to this, the focus of empiricists will lie on how existing mathematics is *actually* (but not-inevitably) rooted in our dealings with the material world, how it has historically evolved, etc. The structure of this paper is as follows: To begin with, in Section 2 the topic of plausible reasoning in mathematics is addressed, through one of its most famous advocates, at least in philosophical circles, George Pólya. After that, we start tackling the issue of mathematical empiricism, in Section 3 in general, then in Section 4 in its more specific, *dialectical* guise, as embodied by Imre Lakatos (who was influenced by Pólya). This will lead us into considerations about the further development of Lakatos's philosophy of mathematics, a topic which will occupy us in Section 5.

2. FROM CERTAINTY TO PLAUSIBILITY

Mainly from reading Descartes' *Regulae ad directionem ingenii* (Rules for the direction of the mind, c. 1628) and Mach's history of mechanics, as well as from a love for Euler's mathematical style, George (born György) Pólya (1887-1985) increasingly grew to appreciate the importance of the context of discovery in mathematics, and thus developed a serious interest in heuristics.[1] As a result, in *How to Solve It* (HSI, 1945),[2] his most famous and accessible work, Pólya set out a four-stage model of the mathematical problem solving process, according to which one (1) first tries to understand any problem at hand, (2) on the basis of that devises a plan, (3) then carries out that plan, and finally (4) returns to the problem and verifies whether it has thereby been satisfactorily settled. Years later, the efforts of this first publication were continued and expanded in two double-volume books, providing a myriad of examples, more sophisticated ones in *Mathematics*

1 Albers and Alexander, 1985, 251.
2 Pólya, 1957.

and Plausible Reasoning (MPR), and more elementary ones in *Mathematical Discovery*.[3]

This extensive body of work has been of profound influence on pedagogy both in America and abroad (the initial book having been translated at least fifteen times), especially in the 1980s. Even at its height, however, the status of Pólya's heuristics remained the subject of much debate and empirical pressure. More specifically, in addition to a minimal amount of clear practical success, the account was particularly suffering the attacks of artificial intelligence (AI), a discipline which seemed to do strikingly better using other, less *fuzzy* approaches. "In essence, the difficulties with the implementation of Pólya's ideas were that (a) they were not specified in adequate detail for implementation, and (b) they appeared to be superseded by more *general* methods".[4] In the last two decades, however, the tables seem to have turned in favor of Pólya, as a consequence of which his spirit definitely lives on in mathematics education.

> First, cognitive science has provided methods for fleshing out the details of Pólya's strategies, making them more accessible for problem solving instruction. [...] In addition, the general methods of AI have turned out to be much weaker than had been thought; methods once thought general and powerful have turned out to have limited scope and power. Research from the past decade [the 1980s] indicates that problem-solving strategies are much more tightly bound to domain-specific subject matter understanding than early AI researchers had claimed.[5]

The pursuit of mathematics education proper is outside the scope of the present paper. Instead, what is of clear interest is the extent to which this field has any actual or potential influence on the philosophy of mathematics. And our current protagonist is relevant at this point, as there is no doubt that this particular connection was in fact sought after by Pólya himself. As will be illustrated below, Imre Lakatos picked up this implicit appeal and endorsed Pólya's ideas, to begin with through the in-depth investigation, in the illustrious study *Proofs and Refutations*, of one of Pólya's specific examples. Let us therefore have a look at where Pólya gets closest to being philosophical himself, which will allow us to elaborate on the role of the non-deductive in mathematics. This is at the outset of MPR's first volume, *Induction and Analogy in Mathematics*, where Pólya himself labels his book as "in a sense" a philosophical essay, and then develops the difference

3 Resp Pólya, 1973, 1968, 1962, 1965.
4 Schoenfeld, 2000, 235.
5 Ibidem.

between *demonstrative* and *plausible* reasoning.[6] The former method, aimed at proving mathematical statements, "has rigid standards, codified and clarified by logic, [. . . it] is safe, beyond controversy, and final".[7] The latter's standards, on the contrary, are "fluid", and directed towards generating *and* supporting conjectures. This type of reasoning, thinks Pólya, is being unjustly neglected in education. "Certainly, let us learn proving, but also *let us learn guessing*".[8] Demonstrative reasoning, he claims, is but one side of mathematics, viz., the format of results, *finished* mathematics. "Yet mathematics in the making resembles any other human knowledge in the making. You have to guess a mathematical theorem before you prove it; you have to guess the idea of the proof before you carry through the details. You have to combine observations and follow analogies; you have to try and try again".[9]

The two kinds of reasoning do not contradict but complement and complete each other. Both are indispensable for any ambitious student of mathematics. From them, plausible, heuristically driven reasoning constitutes *the* way to establishing rigorous proof. Focusing on this hitherto neglected type, Pólya offers many examples; this he considers the most proper manner of instructing the interested reader, offering him or her ample material for imitation and opportunity for practice. There exists, in his view, no foolproof method to learn guessing. Explicit heuristics and search strategies may of course be offered, but these, in principle, will only approximately fit any of the concrete circumstances in which they are later applied. Pólya's approach therefore is, as he himself calls it, naturalist in spirit. "I collect observations, state conclusions, and emphasize points in which my observations seem to support my conclusions. Yet I respect the judgment of the reader and I do not want to force or trick him into adopting my conclusions."[10] Being essentially unable to literally retell the details of the examples offered, he reconstructs their history in a *plausible* way.

Pólya suggests that his results be put to philosophical use by introducing them into the discussion about induction in mathematics—inductive reasoning being a special kind of plausible reasoning. Drawing attention to the case of Goldbach's Conjecture, he aptly points to a lost mathematical tradition (involving, e.g., Euler and Laplace) of acknowledging a vital place for inductive evidence in mathematical inquiry.[11] In the sciences in general,

6 Likewise, in the introduction to HSI, he had distinguished the Euclidean or deductive from the experimental or inductive way of mathematics.

7 Pólya, 1973 (MPR I), v.

8 Pólya, 1973 (MPR I), vi.

9 Pólya, 1973 (MPR I), vi.

10 Pólya, 1968 (MPR II), v.

11 See indeed Echeverría, 1976.

he notes, the proper selection of experimental material is considered to be of utmost importance. For Pólya, mathematics lends itself very well to a study of its inductive patterns of reasoning (i.e., how the degree of *confidence* in a conjecture is influenced either way by the bringing in of evidence), because of the inherent simplicity and clarity of mathematical subjects. Pólya himself, in MPR (II, chs.1-4), actually attempts such an investigation for a variety of specific cases, remaining however very cautious as to its proper philosophical calibre.[12] Also in MPR (I, ch.2),[13] Pólya reflects on the heuristic devices of generalization, specialization, and analogy. Generalization and specialization he considers pretty straightforward and precise techniques often used in mathematical inquiry, viz., when trying to prove a more general proposition than the one at hand, or trying to prove its special cases first. Mathematical analogy, on the other hand, seems more difficult to put one's finger on. Pólya defines it as a type of conceptual similarity essentially rooted in the intentions of the thinker using it. Clarification of analogy thus consists in pinpointing the shared concepts involved.[14]

In §6 of the chapter mentioned, Pólya develops the historical example of Euler's derivation, by analogy, of a formula expressing the sum of the reciprocals of the sequence of squares (1 + 1/4 + 1/9 + 1/16 + ...), viz., $\pi^2/6$. Euler arrived at it by applying a decomposition rule of algebraic equations to an equation which is clearly not of that type, viz., $\sin x = 0$ (being of infinite degree). This was a daring move, but Euler found he had good reason to be confident, as "the numerical value for the sum of the series which he [had] computed before, agreed to the last place with $\pi^2/6$."[15] Notice that these reasons were inductive and not demonstrative, although later Euler was able to actually prove the equality, and thus deductively confirm his informal reasoning. While undoubtedly in favor of the exact mode of mathematical reasoning (like Pólya), "Euler seems to think the same way as reasonable people, scientists or non-scientists, usually think. He seems to accept certain principles: *A conjecture becomes more credible by the verification of any new consequence*. And: *A conjecture becomes more credible if an analogous conjecture becomes more credible*".[16] This being said, induction alone clearly does not suffice. Typically, a (*true*?) mathematician cannot rest

12 "If it is philosophy, it is certainly a pretty low-brow kind of philosophy, more concerned with understanding concrete examples and the concrete behavior of people than with expounding generalities" (Pólya, 1973 (MPR I), viii).

13 A chapter that was reprinted in Tymoczko, 1998, 103-24.

14 E.g., two systems are analogous "if they agree in clearly definable relations of their respective parts" (Pólya, 1973 (MPR I), 13).

15 Pólya, 1973 (MPR I), 20.

16 Pólya, 1973 (MPR I), 22; original emphasis.

content with it, and wants to pass from the inductive into the demonstrative phase of inquiry. "The verification of any consequence increases our confidence in the conjecture, but the verification of the [inductive generalization] can do more: it can *prove* the conjecture."[17] But what if things stop short before formal proof is ever attained? What if there are, at least in certain cases, limits *in principle* to the scope of mathematical (demonstrative) thinking itself? Might it then not be appropriate to actually embrace a *quasi*-empiricist or naturalist philosophy of mathematics more firmly?

We shall attend to the positive answer to the latter question in the next section. For now, let us assess the extent to which Pólya has been of actual influence on this philosophical course. That there is at least an affinity seems clear. Pólya must be considered a frontrunner of quasi-empiricism (or whatever you want to call it), if only because of his implicit recognition of mathematical practice (*in casu* plausible reasoning) as philosophically important. Note that he refrained from ever explicitly proclaiming as much. The final products of mathematics, completed proofs, apparently remained for him beyond controversy, residing as they did in the realm of "demonstrative reasoning," to be dealt with in an isolated context of justification.

> Pólya's work had little impact on the philosophy of mathematics until it was taken up by the more radical quasi-empiricists. They push his analysis one step further by questioning the assumption of completely safe proofs. [. . .] According to [them . . .], it is our informal proofs, the kind investigated by Pólya, that are often safer than any derivative formalizations of them. Once this extra step is taken, Pólya's work ceases to be a mere gloss on the foundational conception of mathematics, but instead becomes a genuine alternative to it.[18]

Pólya exerted most of his philosophical influence in the above sense through Lakatos, who ascribed to him his own interest in mathematical heuristics,[19] and who also derived from Pólya the theme for his most famous work. In MPR (I, Ch.3), Pólya describes Euler's development of the polyhedron conjecture V-E+F=2, without however addressing its proof at all. That is where Lakatos picks it up, to turn it into a detailed historical reconstruction (see Section 4). Lakatos departs from Pólya, who is mainly in the business of identifying "problems ready to be proved" (or refuted), by pointing to the *co-development* of proof and assertion. Although both men

17 Pólya, 1973 (MPR I), 110.
18 Tymoczko, 1998, 97.
19 Particularly HSI, which he translated into Hungarian.

highlight the development of mathematical knowledge, there is thus a big difference. Pólya restricts himself to the context of discovery and embraces the fundamental inductive nature of it. The latter feature was apparently unacceptable to Lakatos.[20]

> Pólya's heuristics may be closer to the students, and that of Lakatos to the working mathematicians. In these cases, Pólya and Lakatos speak of different processes. Pólya speaks of learning mathematics, Lakatos of making mathematics. This is why Lakatos's first rule states: if you have a conjecture, set out to prove it and to disprove it. Both processes belong to scientific research.[21]

3. MATHEMATICS AS REALLY REAL

After this consideration of Pólya and his use of inductive methods of plausible reasoning in mathematics, we now move to a more general philosophical level, viz., that of mathematical empiricism. We will then discuss, in Section 4, one of the most original theorists of mathematics, and who has been explicitly inspired by the pedagogical perspective of Pólya, his fellow-countryman Imre Lakatos. As a starting point, remember that in Kant's view, Euclidean geometry provided us with *a priori* knowledge about actual space. The development of non-Euclidean geometry was a first blow to this account; the availability of full-fledged alternatives allowed the dismissal of any exclusivity claim, at least until empirically verified. In any case, if Euclidean geometry could have then established its syntheticity, i.e., if it indeed had been shown to be a correct description of reality, it would have ceased to be *a priori*, and would have become *a posteriori* instead.

In fact, something worse than that happened: the position of Euclidean geometry was further weakened by Einstein, who in 1916 actually made use of *Riemannian* geometry to arrive—in combination with the equations of his General Relativity Theory—at a cosmological model that *did* receive empirical approval. This not only confirmed the loss of Euclidean privileges as far as representing reality was concerned: the model was exposed as offering a plainly *false* picture of reality (at least on a cosmological scale). In reaction to this geometrical crisis, a turn towards arithmetization occurred in subsequent foundational studies. But what if, instead, mathematicians had rejected such a move as *ad hoc*, and decided rather to bite the bullet. Then empiricism would have perhaps turned out to be a most elegant alternative,

20 Lakatos touches upon this in the case-study referred to Lakatos, 1976, 74.
21 Kiss, 2002, 251.

for therein "the correctness or otherwise of [mathematical] results must be as much an empirical question as that of the correctness or otherwise of a scientific theory."[22] Another possibility, instead of reversing epistemic priority, is to consider geometry and arithmetic as distinct, logically independent disciplines, and to combine a different treatment of both. One might, e.g., become an empiricist in geometry, while safeguarding the *a priori* in arithmetic. While such a peculiar combination seems quite exotic, it is not to be excluded. Indeed, Brouwer and Poincaré might in some sense be considered representatives of this approach. In contrast, the empiricist attitude has almost invariably been defended in geometry and arithmetic alike (whether or not reducing one to the other).

Scientists, in general, are in the business of describing and explaining the world as it appears to us. This necessarily includes some kind of grasp of the mechanisms at work beyond the *merely apparent*—i.e., what is immediately accessible via the senses only. Especially over the past few centuries, hand in hand with conceptual evolution, the invention and rising accuracy of all sorts of instruments have brought about an enormous improvement of both qualitative observation (e.g., the microscope and the telescope) and quantitative measurement (e.g., of temperature, pressure, weight, length, size, etc). Although for many, science has remained synonymous with *natural* or *exact* science, a recognition of the variety and complexity of phenomena to be systematically dealt with has resulted, at present, in dozens—if not hundreds—of disciplines and subdisciplines operating under its wings, including *human* and *social* sciences. While the latter's object of study, human behavior (individually and collectively), belongs to the spatio-temporal world as much as trees or planets do, its operating mechanisms remain *largely* concealed, buried deep down within such elusive things as the mind, culture, or society, there being far from even a beginning of consensus regarding the instruments and methods of properly charting them.

Nevertheless, postulating an essential divide between the *rock-certainty* of the natural sciences and the speculative *loose talk* of humanities and sociology, is drawing too harsh a distinction. Recurring philosophical controversy about this matter, though seemingly fought over purely intellectual values, particularly Holy Truthfulness, also to a large extent thrives on *strategic* conflicts of interest regarding interconnected issues such as demonstrable usefulness (economical or other), appropriate recognition thereof, and—consequently—proper funding. While on the face of it this topic seems rather remote from any discussion of the nature of mathematics, it is nevertheless of importance in order to appreciate the distinction between two types of empiricism in the philosophy of mathematics. The original

[22] Tiles, 1996, 339.

conception, physicalism, to be discussed in the rest of this section, is about one and a half centuries old; the other, dialectalism, the topic of the section, is of more recent date.

Physicalism, quite unsurprisingly, is the *meta*physical theory that all is physical. As a philosophy of science, it holds that all meaningful discourse is expressible in the language of physics. The origin of the latter discipline lies in the materialistic tradition of the *natural philosophers* in ancient Greece, who replaced animist explanations of the world by theories describing observable, material, or natural processes. In its modern sense, physics is the result of a nineteenth century synthesis of some then existing disciplines, such as mechanics, optics, and electricity. Physics is the basic of the physical sciences, probing for the most fundamental laws of nature. Physicalism about mathematics, as pioneered by the father of utilitarianism, John Stuart Mill (1806-1873), implies that mathematics, the *queen* of the sciences,[23] is considered an empirical science among all others—i.e., based on deduction from premises that have first been inductively established. Its particular task is capturing the most general properties of the world. Mathematical objects, then, at least the fundamental ones,[24] are not real in any abstract sense, but *really* real, that is, in the most concrete of ways: they exist inside space-time, constituting properties of and relations between material things.[25] Consequently, physicalists denounce not just Platonism, but also two other popular positions: constructivism and conventionalism, as all of these defend a form of non-empirical or *a priori* access to, or at least characterization of, true mathematical knowledge (as explained very briefly in the introduction).

Like Frege, although one of his fiercest opponents,[26] Mill thought of conceptualism as confusing logic with psychology—i.e., propositions with judgments, and properties with ideas. The human mind, for him, could not be the innate seat of mathematical truth, as it was but another piece of nature. In contrast to versions of nominalism, he, like Hobbes, maintained the identificalness connotation and denotation of names. At the risk of being but a circular play-on-words, Mill believed, mathematics cannot be merely analytic: numerals *connote* attributes of aggregates while *denoting* the aggregates themselves, thus actually saying something about the world. Hence, for Mill, the *real* basis of mathematical knowledge was inference

23 Such Gauss is reputed to have first called it. See Bell, 1938.

24 Indeed Mill seems to have made the explicit distinction between "contingent" inductive generalizations and "necessary" deductive reasoning on the basis of them (Van Bendegem, 1990).

25 Bigelow, 1988, 1.

26 In Frege's *Grundlagen* (§7 and §23) is contained a bitter critique of the physical underpinning of arithmetic propagated by Mill. For a commentary see, e.g., Bigelow, 1988 (§4) or Resnik, 1980 (Ch.4).

from observation, i.e., generalization, up to and including idealization. "In this sense, the propositions of geometry are inductive generalizations about possible physical figures drawn in physical space. They have been confirmed by long-standing experience."[27] Similarly, in arithmetic, numerals are general terms ranging over aggregates of ordinary objects.[28] These are not unproblematic suppositions. Exactly how are the ideal mathematical objects, e.g., perfect figures and large numbers, to be derived from inherently limited human experience? The Millian line here seems to be that they aren't, which brings out a very important facet of empiricism, when consequently thought through: fallibilism. This term, alleged to have been first coined by Charles Sanders Peirce (1839-1914),[29] points at the circumstance that, even in science, we cannot be absolutely sure of anything, but rather can have only approximate knowledge. Why does mathematics nevertheless strike us as *a priori* and universally true? The empiricist's answer is Humean: through repeated association of what co-emerges, we can hardly conceive things to be otherwise.

Another related weakness of Mill's account is apparently its limited scope. Because of a restriction as to what real human agents are capable of actually performing, "Mill only deals with geometry, arithmetic [little more than simple sums and differences, or what is learned in elementary school], and some algebra, not the branches of higher mathematics".[30] Present-day empiricists like Philip Kitcher or John Bigelow, therefore, are surely more liberal when it comes to characterizing idealization. "Instead of speaking of the collecting and constructing activities of *actual* humans, Kitcher speaks of the activities of *fictitious* constructors who do not share human limitations of time, space, attention-span, or even lifetime."[31] Bigelow, from his side, has developed a physicalist account of mathematics (and numbers in particular) as the theory of *multiply located universals*, extending at least over all *primary* qualities of material objects.[32] The higher degree of contemporary sophistication is definitely connected with the considerable attention that is devoted today to the place and function of mathematics in the whole of scientific practice. Indeed, to consider mathematics as a science (good or bad) like the others, as naturalists do, should not imply epistemologically

[27] Shapiro, 2000, 94. For specific details on Mill's conception of geometry, see Torretti, 1984 (§4.1.1).

[28] See Kitcher, 1995 for an analysis.

[29] Peirce seems to have held a rather dubious position as far as mathematics was concerned, though. See, e.g., Goudge, 1969, 39-41, or Cooke, 2003.

[30] Shapiro, 2000, 100.

[31] Shapiro, 2000, 101; emphasis added. See Kitcher, 1983.

[32] As distinguished by John Locke in *An Essay Concerning Human Understanding* (1689), these include solidity, extension, figure, motion or rest, and number (Bigelow, 1988, 14).

and/or methodologically disconnecting it from these, as one particular naturalist (Maddy[33]) claims. After all, mathematical language is amply and ever increasingly used in the other sciences, natural, life, human, or social ones alike.[34]

4. QUASI-EMPIRICISM OR DIALECTALISM

The conception of *quasi*-empiricism in mathematics is usually ascribed to the Hungarian Imre Lakatos, born Lipschitz (1922-1974).[35] Having served time, first in the Communist government, and then in a Communist jail, in 1956 Lakatos fled to the West. After some years in Cambridge, in 1960 he became a lecturer at the London School of Economics, where he was associated with Karl Popper (1902-1994). Lakatos's philosophy of mathematics is inextricably connected with his casestudy *Proofs and Refutations: the Logic of Mathematical Discovery* (PR; 1963-4), a series of papers extracted from his doctoral dissertation, edited and published posthumously in one volume in 1976.[36] Its topic is Euler's conjecture V-E+F=2, the respective variables standing for the number of Vertices, Edges, and Faces of any polyhedron. In this section, we shall largely limit ourselves to commenting on the significance of this exercise in *rational reconstruction*, which reformulates the historical narrative in order to allow it to be tested against a proposed logical scheme of scientific progress. As the actual history therein gets relegated largely to the footnotes, Lakatos's

33 See e.g., Maddy, 1997.

34 Besides other sources mentioned, this section has substantially benefitted from Encyclopedia Britannica CD Multimedia Edition, 1999 (entry "Natural Sciences"); and Routledge Encyclopedia of Philosophy, Version 1.0, London: Routledge, 1998 (entry "Mill, John Stuart" by John Skorupski). For an elaborate review of some recent empiricists' work, see Milne, 1994.

35 Oliveri, 1997 (§2) includes the Kantian and Quinean conceptions as being quasi-empiricist in nature, meaning that they both strike a balance between the rational or *a priori* and the empirical or *a posteriori*. Although Lakatos has arguably drawn on some of Quine's ideas (such as holism or underdetermination), none of these compromises shares with Lakatos's version the feature of fallibility, a result of accepting mathematics as a science among the others, where what is considered truth and falsity evolves over time. Therefore, Oliveri proposes a definition of quasi-empiricism encompassing all these versions: "a mathematical theory is quasi-empirical just in case experience is indispensable to provide an account of the knowledge produced by it even though the statements of the theory are *a priori* true" (op.cit., 232). A more restrictive definition (op.cit., 234) then corresponds to the Lakatosian account, just leaving those theories quasi-empirical that are actually falsifiable (viz., in the sophisticated manner of Lakatos, 1972; see Section 5).

36 Lakatos, 1976.

format has also disapprovingly been called *footnote history*. Below, in Section 5, we shall turn to a more thorough discussion of Lakatos's rationalism about scientific progress. Many a present-day mathematical naturalist or humanist has been inspired either directly or indirectly by the Lakatosian body of thought. Therefore some core naturalist interests are foreshadowed here, most notably and generally, the primacy of mathematical practice over philosophical discourse about it. It is important to note right away, however, that Lakatos never developed a full-fledged philosophical program for mathematics himself. The simple reason for this is as mundane as can be: his untimely and sudden death. Although after PR he left the subject, to embrace more generally scientific concerns, there is clear evidence that he *did* plan to return to it in a systematic way.

What makes Lakatos's approach to mathematics, provisional as it might be, particularly salient, is its combination—or synthesis, if you will—of three major influences exerted on him: Hegel's dialectics (mainly through George Lukács, back in Hungary), Popper's falsificationism (in London), and Pólya's plausibilism, highlighted in Section 2.[37] The latter mathematician, who was largely responsible for the outspoken heuristical dimension to Lakatos's approach, unwittingly suggested through his writings the topic for PR by describing Euler's inductive derivation of the conjecture mentioned above. Lakatos picked it up there, and went on to look into the proof process as well (completing the casestudy, as it were). More generally, Pólya inspired Lakatos's philosophical inclination to pay attention to the context of discovery. Indeed, for Lakatos, the proper way to evaluate mathematical heuristics—i.e., all the instruments put to use on the road to mathematical conjecture and proof, was through dialectics. As Lakatos confided in a letter to Max Wartowsky, he had the solemn wish to establish a well-founded dialectical school in the philosophy of mathematics.[38] While in non-dialectical logic attention remains restricted to static patterns of inference between propositions, students of the dialectical are also interested in conceptual dynamics. Accordingly, a dialectical philosophy of mathematics looks into what drives the argumentative development of mathematical concepts. PR prominently exhibits this in its dialogue form (a technique also forcefully exploited before by famous philosophers such as Plato, Berkeley, or Hume), staging the search for a proof for Euler's formula as a classroom discussion between various (invariably brilliant) students representing

[37] This triple indebtedness was pointed out by Lakatos himself at the outset of his Ph.D., part of which would evolve into PR (Ernest, 1997, 117n2).

[38] See Larvor, 1998, or 2001, 9 resp 212.

different positions in the debate.[39] In the course of this discussion, two modes of conceptual refinement surface (specifically in the definition of *polyhedron*): the presenting of a novel kind of geometrical object (involving the famous *monster-barring*, or dismissing certain constructions as genuine polyhedra),[40] and the proposing of a novel kind of proof (involving new techniques that allow the incorporation of *local counterexamples* or *guilty lemmas* into the conjecture as conditions).

The point about the fallibility of mathematics as an epistemological endeavor connects Lakatos's account with his London teacher and colleague Popper and with empiricism, the general topic of the previous section. Lakatos explicitly tackled this topic in the paper *A Renaissance of Empiricism in the Recent Philosophy of Mathematics* (1967),[41] in which he affirmed not that mathematics is *exactly* like the rest of *a posteriori* science (i.e., the rarely defended Millian position), but that it is in any case *much* like it. More particularly, it is liable to the Popperian upward flow of falsity rather than to the Euclidean downward flow of truth (as defended by the big foundational programs). This in order to accord it with the more refined—Hegelian—dialectical mechanism of informal negotiation, a mechanism which involves—Pólyan—heuristic falsifiers (e.g., lack of generality or explanatory power),[42] as described in PR. Consequently, for Lakatos, mathematics is a quasi-empirical science. Popper's influence on PR is clearly acknowledged by the work's title and subtitle, which straightforwardly suggest an application to mathematics of both Popper's *Conjectures and Refutations* (1963) and *The Logic of Scientific Discovery* (1959).[43] In these works Popper first set out his general method of conjectures and refutations, depicting good science as a maximally vulnerable thus fallible enterprise, open to disconfirmation at any point and time. In other words, potential falsification is more important than is

39 "Alpha is a constant creator of counterexamples to Euler's Theorem; Beta and Sigma represent methodological aspects of both Cauchy's and Abel's mathematics, a method Lakatos calls 'exception-barring', and which plays, he argues, a key role in the mathematics of the nineteenth century; Delta is a 'monster-barrer' who repeatedly contracts and redefines his definitions of polyhedra when faced with Alpha's counterexamples; Lambda is the spokesman for lemma-incorporation; and so on" (Kadvany, 1995, 265).

40 'Exception-barring' is a variant whereby polyhedra, although accepted as genuine, are excluded from the field of application of the theorem (e.g., concave ones).

41 Lakatos, 1998.

42 For the latter topic, see, e.g., Sandborg, 1998, and Mancosu, 2000, commenting and expanding on the few earlier philosophical treatments (involving Steiner, Kitcher, and Resnik).

43 Popper, 1978, and 2002.

verification or, in the case of mathematics, proof.[44]

As a result of all this, Lakatos's approach can be rightfully considered as a synthesis of at least these three complementary essential influences: fallibilism (Popper), methodology (Pólya), and intersubjectivity (Hegel). In his own words: "Its modest aim is to elaborate the point that informal, quasi-empirical, mathematics does not grow through a monotonous increase of the number of indubitably established theorems but through the incessant improvement of guesses by speculation and criticism, by the logic of proofs and refutations."[45] Indeed, an important ingredient is the bringing to the fore of aspects belonging to what we introduced above as the context of mathematical discovery, involving the *informal* or productive phases of doing mathematics, which precede its formal or evaluative stage (i.e., the context of justification). Note that for Lakatos, it is perfectly possible to speak of progress in this context. In other words, it is liable to rational philosophical inquiry, and is not to be left to "relativist" treatments by psychologists and/or sociologists. "An investigation of *informal* mathematics will yield a rich situational logic for working mathematicians, a situational logic which is neither mechanical nor irrational, but which cannot be recognized and still less, stimulated, by the formalist philosophy."[46] That is, for formalism, informal mathematics, being neither tautological nor empirical, is meaningless. "Formalism [the obvious target of PR] disconnects the history of mathematics from the philosophy of mathematics, since, according to the formalist conception of mathematics, there is no history of mathematics proper. [. . .] Under the present dominance of formalism, one is tempted to paraphrase Kant: the history of mathematics, lacking the guidance of philosophy, has become *blind*, while the philosophy of mathematics, turning its back on the most intriguing phenomena in the history of mathematics, has become *empty*."[47] Through the latter famous adaption of Kant, "Lakatos indicated that the classical theories in the philosophy of mathematics are not adequate with respect to actual research practice and proposed a new model being closer to that practice."[48] This touches upon a very important point. Although Lakatos's philosophy of

[44] No picture of Popper's own philosophy of mathematics is aimed at here. See, e.g., Glas, 2001a and 2001b. Glas argues, among other things, that there is certainly more to Popper's philosophy of mathematics than its merely having served as an inspiration for Lakatos, who transposed Popper's fallibility thesis from science to mathematics. Particularly, Glas claims that despite his "third world" theory Popper already was a mathematical fallibilist himself.

[45] Lakatos, 1976, 5.

[46] Lakatos, 1976, 4.

[47] Lakatos, 1976, 1-2; original emphasis.

[48] Murawski, 1999, 18.

mathematics is generally identified with or reduced to fallibilism, there is something much more fundamental about it, viz., its "having turned philosophical attention to what one might call the *inner life* of mathematics,"[49] —mathematical practice.

5. LAKATOSIAN RESEARCH PROGRAMS

In the previous section, we introduced Lakatos's *Proofs and Refutations* (PR), leading mathematical theories from the initial phase of naive conjecturing, through various stages of proof re-examination, finally to yield complex, *proof-generated* theorems.[50] As was indicated there, this was not at all supposed to be the end of it; for Lakatos, by the time of his premature death, still dreamt of founding a proper dialectical school in the philosophy of mathematics. In it, Lakatos would have probably wished to distinguish a further phase in the development of mathematical theories, viz., that of their maturity, past the seemingly directionless PR-scheme. That is, as he had done for science in general (apparently under the influence of Kuhn), he would have defined research programs, embodying a particular logic of mathematical discovery—i.e., mathematical methodology put on rational footings. Following Pólya but *contra* Popper, Lakatos held that both justification and discovery are liable to logical investigation,[51] while following Popper but *contra* Pólya, he stripped discovery of its strictly inductive character at the conjectural phase, and was concerned also with the *deductive* flow of truth (or rather falsity) through constructing actual proofs. Further developing this discovery phase as rational could keep fallibilism—the presumed primacy of the informal—from lapsing into skepticism, or Kuhnian relativist *excesses*. Still hypothetically speaking, this is also where Lakatos would have been able to firmly couple his criticism of the foundationalist approach with having mathematics epistemologically fall into place as a non-exceptional part of the general scientific endeavor.[52]

Lakatosian research programs, as conceived in the lengthy paper *Falsification and the Methodology of Scientific Research Programmes*

[49] Larvor, 2001, 213.

[50] For an overview of the seven different stages, see Lakatos, 1976, 127-8. For some critical appraisals, see Currie, 1979, Feferman, 1998, or Steiner, 1983.

[51] "[H]is research demonstrates convincingly that mathematical discovery not only belongs to the field of psychological analysis but also can be analysed rationally" (Yuxin, 1990, 382).

[52] Yuxin, 1990, 388.

(MSRP; 1970),[53] consist of a *hard core* of laws never to be given up without dissolving the program, and a *protective belt* of laws which on the contrary *can* be fiddled with in the light of recalcitrant experience. Also adhering to a research program, are negative and positive heuristics, together constituting its inherent methodology: keeping the hard core from being affected, and driving the protective belt to pro-actively develop itself (i.e., heuristic progress), respectively.[54] The result has been called a synthesis of Popper's and Kuhn's philosophies of science, but also clearly steers away from both. As opposed to Popperian "naive" falsification, theories are not abandoned just like that. They are part of a historical chain situated within a more sophisticated program that is engaged in a competition with rival ones. For Popper no alternatives need be available in order to refute theories. As opposed to Kuhnian revolutions or *Gestalt*-switches between successive incommensurable paradigms, then, past the pre-scientific stage, old research programs are not necessarily overturned by new ones, but may continue to exist next to these, even when (apparently) superseded.[55] This points to a moderately rationalistic caliber: although there are normative reasons for preferring a paradigm (which means they are in principle comparable), not all scholars will necessarily agree on what criteria are in play and/or dominantly important; this mitigates the rational element, allowing for psychological and/or sociological influences, without however lapsing into total irrationality. Also, with regard to mathematics, "MSRP was [the later] Lakatos's answer to the extreme sceptics. Rational reconstructions are Lakatos's main tool to defend both the fallibility of mathematical knowledge and the rationality of its development."[56]

A number of elaborations of MSRP into the field of mathematics have been attempted, e.g., by Hallett, Glas, and Koetsier.[57] We shall have a combined look into the latter two. As a starting point, it should be understood that Lakatos saw a continuity between PR and MSRP, as he saw isolated theories gradually turn into mathematical research programs. Basic to this process are the lemmas (including conjectures) that are not refuted in the course of applying the method of proofs and refutations. They develop

[53] Lakatos, 1972, originally published in the volume ensuing from the Popper-Kuhn debate organized by Lakatos in the late 1960s, and a modified version of the 1968 paper of this title published in *Proceedings of the Aristotelian Society* 69: 149-86.

[54] Lakatos considers Newton's celestial mechanics as exemplary for his proposed methodology.

[55] Koetsier, 2002, (§4.2) develops the mathematical example of the major rivaling foundational programs of the early twentieth century: logicism, intuitionism, and formalism.

[56] Koetsier, 2002, 189.

[57] Hallett, 1979a and, 1979b, Glas, 1995, Koetsier, 1991.

into axioms, to form the hard core of the research program, to which no straightforward contradiction is any longer possible. Through mechanisms forming the so-called *negative* heuristic, any topic for discussion is relegated to the protective belt of auxiliary hypotheses and concepts. On the other hand, the method of concept stretching induced by producing counter-examples is a PR-forerunner of the "positive" heuristic in MSRP.[58] This continuity claim however raises a difficult issue: that of the level at which research programs operate. Lakatos himself apparently wavered between situating them at a higher-order level, e.g., like Kuhn, that of the discipline (where the central MSRP concepts are however extremely hard to operationalize), or rather at some lower level in between that of research branch and individual theories, viz., consisting of a cluster of the latter (which seems more plausible).[59] Whatever Lakatos's intentions, it seems clear that actual patterns of development are more diverse and intricate than would allow for a simple choice between these two levels (which would perhaps suggest a pluralistic stance).

In relation to this, Glas considers the main reason MSRP resists straightforward application, especially to mathematics, is that it is too restrictive, and leaves a vital discrepancy between actual history and a rationally reconstructed one.[60] In the light of various cases developed by him, Glas maintains that theories do not develop as autonomously as Lakatos would have us believe, and that external factors are more important than is allowed for by his framework.

> Lakatos's methodology is aimed, not at historiographic explanations of scientific development, but at the rational reconstruction of what is considered to be its [Popperian] third-world equivalent, to which alone methodological standards apply. Methodology has itself to be appraised by the extent to which it is able to perform this task without having to appeal to factors that are *external* to it, hence *non-rational* by its intrinsic methodological standards. *Internal* history in this—unusual—sense is said to be self-sufficient, and *external* history irrelevant for understanding the growth of disembodied objective knowledge. The Lakatosian reconstruction of the history of science and mathematics presupposes that its *rational* aspect is fully accountable by the *logic* of discovery, whereas external factors are invoked only for the non-rational explanation of the

58 Glas, 1995 (§1), Yuxin, 1990 (§2.2).

59 Kvasz, 2002 (§3.1).

60 Glas, 1995.

> remaining discrepancies between real history and its rationally reconstructed logical substitute.[61]

Although Lakatos, in translating Kuhn's findings, *de*revolutionizes the latter account, assimilating problems by patching up existing programs rather than allowing crises to erupt (which in the case of mathematics makes no difference anyway, as Kuhn was a disbeliever in this), we find this accusation of implicit internalism and weak sociology, thus concealed apriorism, a bit exaggerated. Conversely, we think Lakatos's framework, when elaborated, does offer opportunities for a genuine naturalist philosophical treatment, potentially offering a well-balanced place to both empirical findings about mathematical practice and its epistemological assessment, thus steering between both extremes of externalist contingency and internalist normativity. Of course Lakatos should then be taken as an important source of inspiration, calling for an abandonment of linear and cumulative historiography. This other, more favorable approach is not incompatible with a critical one. This is even suggested by Glas himself. He notes that, while perhaps working very well *within* research, MSRP fails to account for the so-called rational comparison *across* research programs, which leads to their overtaking one another. This would be particularly the case for criteria concerning the social (e.g., educational) functionality of mathematical programs. Thus factors seem to be in play that cannot be accounted for from within Lakatos's *internalistic* framework. Yet, as Glas himself recognizes,[62] this was implicitly admitted by Lakatos. "So the Lakatosian methodological perspective must at least be complemented by a socio-historical perspective in the Kuhnian vein in order to account for the latter sort of changes [i.e., meta-level innovations drawing upon different programs]."[63] The seemingly strict separation between the rational and the social awaits annihilation (as, for another also does Kuhn's separation between the locally and the globally social).

Clearly that young mathematicians do not enter the immense field of inquiry just like that, but rather choose a particular segment of it. Therefore, some additional structure seems to be called for on top of Lakatos's (and Pólya's) low-level heuristics of proving and refuting. To begin to accommodate threatening confusion and/or deadlock with respect to MSRP, as far as this larger structure is concerned, Teun Koetsier has proposed an alternative to the later Lakatos account: distinguishing, instead of programs, research *projects* and *traditions*, a kind of enriching differentiation which we

61 Glas, 1995, 234; original emphasis.
62 Glas, 1995, 239.
63 Glas, 1995, 242.

think—at least in principle—should be applauded from a descriptive point of view.[64] A research tradition, operating on the macro level, carries *general* assumptions about the appropriate objects and methods of study; while a research project, operating on an intermediate level, sets a *specific* agenda of goals and tools.[65] On both levels, a methodology of mathematical research traditions is at work, driving progress through the favoring of traditions and projects that generate important and likely-to-be-solved conjectures, and then going on to actually succeed in proving or refuting these. Some examples will be given below.

As he concedes, Koetsier is indebted to a large extent to the ideas of philosopher of science Larry Laudan, a leading figure, with Lakatos, in the kind of approach which blends, in the wake of Kuhn, historicity with rationality. Because of their clinging to the latter, and in view of their avoidance of reflexivity (i.e., the application of a proposed methodology to one's own theorizing), both Laudan and Lakatos might still be considered by naturalistic purists (or extremists), *proto*-naturalists.[66] On the other hand, both, in their account of scientific practice, are trying to strike a balance between the normative and the descriptive, a quite reasonable endeavor, in our estimate. In putting forward specific criteria for measuring scientific rationality or progress, they propose *meta*methodologies (a term coined by Lakatos), that is, epistemologies of methodology. For Laudan, the criterion of whether to *pursue* (as distinguished from *accept*) a specific research program is its problem-solving effectiveness, as determined by a weighted comparison of solved vs. outstanding problems in the discipline under consideration.[67]

Laudan has called his scientific epistemology *normative naturalism*, and has contrasted it to Lakatos's *historicism*.[68] Science, following the latter, embodies its own rationality through the actual methodologies applied in subsequent stages of inquiry, this process being philosophically reconstructible independent of any prior theory of rationality. This indeed turns out to be quite in line with Kuhn's proposal, at least to the extent that the history of science is self-justifying, and thus provides test cases for any supplementary philosophical theorizing. Where (Kuhn and Lakatos) diverge is on whether or not this history is a rational or progressive one. From the early Kuhn, one might go one of two ways: embracing relativism, a line

64 Koetsier's approach has been vehemently dismissed by Glas for being philosophically uninformed, viz., misrepresenting Lakatos (see, e.g., Glas, 1995, 233n5). We have opted to abstain from any comments on this controversy here.

65 Koetsier, 1991 (Ch.VI).

66 Giere, 1985.

67 This project was initiated in Laudan, 1977.

68 Laudan , 1987.

suggested by the early Kuhn (and fully endorsed by Paul Feyerabend, among others),[69] or seeing regular restrictions as being implicitly imposed on past practice and therefore being of interest to future practice, an idea which is more compatible with the later Kuhn's partial return to the *a priori*, and which would then roughly correspond to the option followed by Lakatos, in devising his MSRP.

Normative naturalism, in contrast, "is a prescriptive enterprise whose acknowledged aim is to uncover standards for the appraisal of scientific theories and explanations, and to specify conditions under which theory-replacement constitutes progress."[70] The normative naturalist, following the author of this characterization, John Losee, holds that such standards, first, arise from within the practice of science itself; and second, have provisional status only, depending on future developments in the field (included in this school are Otto Neurath and Philip Kitcher). Returning to Koetsier then, his embracing of rationality is indeed not so much reconstructionist as Laudanesque, meaning that what is rational is understood in terms of what induces actual progress (i.e., concrete problem-solving effectiveness), instead of the other way round.

As far as major scientific shifts are concerned, ever since Kuhn, the mere mention of research traditions (or paradigms, or the like) has invariably raised questions about alleged revolutions. The nice thing about Koetsier's approach is that, unlike Lakatos, it remains impartial on this issue, as progress as such is its crucial point, whether revolutionary or not. In his book, Koetsier has himself worked out some historical case studies on transitions between research traditions, both from ancient Greek and modern mathematics. Various transitional episodes are accounted for: in Greece, from a practical, demonstrative tradition to the axiomatic, Euclidean one; in modern times, from the latter one to seventeenth-century geometrical, eighteenth-century formalist, and nineteenth-century conceptual ones in the calculus. In the first half of the twentieth century, the most important research tradition is claimed to have been the Bourbaki structuralist one. Examples of research projects (explicitly likened to Kuhn's paradigms as exemplars) include Lagrange's and Weierstrass's analytic schools, part of, or even dominating, the formalist and conceptualist traditions referred to above; and more recently, Gorenstein's classification project, revolving around finite simple groups.

In view of present-day mathematics (a topic largely neglected by

[69] Feyerabend, 1978. "Feyerabend maintained that there is no reason for a practising scientist to consult the philosophy of science. There is nothing in the philosophy of science which can help him solve his problems" (Losee, 2001, 189).

[70] Losee, 2004, 130.

philosophers), David Corfield—who in his turn has put Lakatos to the test—judges that neither Glas (calling to externalize MSRP) nor Koetsier (complicating it) have fully succeeded in their reappraisal of Lakatos's account. The former does not consider the higher-level interaction between different programs and also seems to put a rather unilateral stress on outside determinants; while for Corfield the ludic character of mathematics, i.e., mathematicians fooling around without any external concerns, should not be neglected. Instead, Corfield urges us to simultaneously appreciate internal and external factors and also attend to the interweaving of theories, when devising a theory of mathematical practice.[71] Koetsier, in his turn, although being applauded for proposing a more complex methodological schema, is said not to have pushed this exercise far enough.[72] For one thing, his formulation in terms of conjectures and theorems seems to have become outdated by now. "Other candidates for signs of progress include the reorganization of existing bodies of work and the production of new techniques to solve problems which need not be theorems."[73] As for the classification with traditions and projects, variation is said to be still insufficient. For example, for the twentieth century, structuralism is presented as virtually the one remaining tradition, "and so the term 'research project' is left to do an immense amount of work. In effect it has to describe all the developments that have occurred since, say, the 1930s. [. . .] Why not then allow for three levels—tradition, program, and project—each of which may fade away through lack of results or be superseded or swallowed up by a rival, the latter two also being able to spur new developments in the next higher level."[74] Corfield gives the forceful example of Category Theory, which "began as a project to study continuous mapping within the program of algebraic topology, has since become a program in its own right, and is challenging set theory to become the language of the dominant tradition".[75]

As for the original Lakatosian MSRP, Corfield, in view of the developments in algebraic topology, i.e., the field of Euler's conjecture treated by Lakatos in PR, deems it extremely hard to hold on to, especially its conception of a hard core of immutable beliefs.[76] The reason is that so-called homology theories were fairly easily admitted into the field, despite their being in utter contradiction with its basic axioms. Therefore, Corfield suggests, instead of maintaining a strict policy of admission to the hard core

71 Corfield, 2003 (§8.1).
72 Corfield, 2003 (§8.5).
73 Corfield, 2003, 197.
74 Corfield, 2003, 198.
75 Ibidem.
76 Corfield, 2003 (§8.2).

(viz., only allowing axioms), we ought to move towards a wider notion, and thus pass "from seeing a mathematical theory as a collection of statements making truth claims, to seeing it as the clarification and elaboration of certain central ideas by providing definitions to isolate classes of relevant entities and ways of categorizing and organizing information about entities."[77] This proposed shift involves an important role for the concepts of *model* or *construction*, in the sense that what is of primary importance is not so much whether a particular formal system can be established as true or false but what a system contributes to providing interesting models or valuable constructions. Through loosening the conception of the hard core, for Corfield, room could also be made for taking into consideration *metaphysical* or higher-level beliefs and/or aims (such as understanding), thus mutually approaching hard core and heuristics. After all, he remarks, "not all research programs are launched by the progressive understanding generated by improved attempts at solving a single conjecture."[78] Some, that is, are to serve more general purposes, such as refining the conceptual apparatus or deepening existing theoretical insights.[79]

6. CONCLUSION

Let us sum up. Lakatos attacked the logical approach to mathematics, which has been entirely focused on laying bare the time- and culture-independent, rational nature of the enterprise. The problem is that if such a stance were really to provide a truthful picture of current (or even ideal) affairs, then it would become simply impossible to account for various past (or alternative) methods, as these would be deemed utterly irrational from this *presentist* perspective. Lakatos proposed to solve this tension by philosophically embracing the historicity of mathematics—conceptual development, as part and parcel of genuine practice. Instead of mathematics being cumulative in its dynamics, room was now being made for error and improvement, through the rational dialectics of PR. In particular, although mathematical knowledge cannot be refuted in view of recalcitrant experience, it can be informally attacked and thus driven to further refinement or progress (improvement of proofs and techniques). Lakatos later developed a framework suited to account for this circumstance: MSRP. As applied to mathematics (which it never was, at least by him), it would ideally have been "intended to be

77 Corfield, 2003, 181.
78 Corfield, 2003, 182.
79 General sources appealed to for this section: Corfield, 2003 (Chs.7-8), Yuxin, 1990.

normative of good mathematical practice,"[80] viz., able to explain or rationally reconstruct more of the actual history of mathematics than any of its rivals. Lakatos did not bring about a radical break with past logical approaches, however; for *methodically* reconstructed, not actual, history was his testing ground, thereby *explaining* real history. On the other hand, he did turn out to be a major source of inspiration for philosophers to embrace socio-historically tinted approaches to mathematics, and thus is clearly one of the major initiators of the externalist tradition that has philosophically translated the Kuhnian principles to mathematics, this former "paragon of certitude" (Hilbert).

In conclusion, one might be left with the impression of being stuck in the middle with Lakatos. "To some [. . .], Lakatos did not go far enough because he only partially admitted social, cultural, and other contingent factors into his philosophy of mathematics. Others would view his halting at the boundary of rationality as a strength, thereby avoiding a slide into sociologism."[81] Since fallibilism was the most distinctive feature of his philosophy of mathematics, at the very least "Lakatos might be considered to have potentially freed the philosophy of mathematics to reconsider its function, as well as to question the hitherto unchallenged status of mathematical truth."[82] It is then up to his students to develop and adapt his uncompleted views to better fit present-day meta-scientific insights. With Glas, Koetsier, and Corfield, we have briefly pointed out some possible directions. In any case, to reconnect with the original issue of mathematical education, we hope to have shown what was, after all, at stake: viz., that at the very basis of Lakatos's quasi-empiricist philosophical excavations lay a central concern, shown through his affinities with Pólya, for the pedagogical principles surrounding the informal phases of mathematical inquiry (heuristics). This suggests that the development of humanistic mathematics is indeed not at all unconnected from the educational dimension—i.e., from teaching and learning it.

REFERENCES

Albers, Donald J., and Alexanderson, Gerald L. (Eds.) (1985). *Mathematical People. Profiles and Interviews*. Birkhäuser, Boston.

Bell, E.T. (1938). *The Queen of the Sciences*. G.E. Stechert & Co, New York (Reprint of the 1931 original.)

[80] Glas, 1988, 22.
[81] Ernest, 1997, 122-3.
[82] Ernest, 1997, 123.

Bigelow, John (1988). *The Reality of Numbers. A Physicalist's Philosophy of Mathematics.* Clarendon Press, Oxford.

Cooke, Elizabeth F. (2003). "Peirce, Fallibilism, and the Science of Mathematics". *Philosophia Mathematica* III, 11 (2), 158-75.

Corfield, David (2003). *Towards a Philosophy of Real Mathematics.* Cambridge University Press, Cambridge - New York.

Currie, Gregory (1979). "Lakatos's Philosophy of Mathematics". *Synthese*, 42 (2), 335-51.

Echeverria, Javier (1996). "Empirical Methods in Mathematics. A Case-Study: Goldbach's Conjecture". In Gonzalo Munévar (Ed.) *Spanish Studies in the Philosophy of Science*, Kluwer Academic Publishers, Dordrecht - Boston – London, 19-55.

Ernest, Paul (1997). "The Legacy of Lakatos: Reconceptualising the Philosophy of Mathematics". *Philosophia Mathematica* (III), 5 (2), 116-34.

Feferman, Solomon (1998). "The Logic of Mathematical Discovery vs. the Logical Structure of Mathematics". In his *In the Light of Logic*, Oxford University Press, New York - Oxford, 77-93.

Feyerabend, Paul (1978). *Against Method.* Verso, London (Originally published by NLB in 1975.)

Giere, Ronald N. (1985). "Philosophy of Science Naturalized". *Philosophy of Science*, 52, 331-56.

Glas, Eduard (1988). "Between Form and Function. Social Issues in Mathematical Change". *Philosophica*, 42 (2), 21-41.

Glas, Eduard (1995). "Kuhn, Lakatos and the Image of Mathematics". *Philosophia Mathematica* (III), 3 (3), 225-47.

Glas, Eduard (2001a). "The 'Popperian Programme' and Mathematics. Part I: The Fallibilist Logic of Mathematical Discovery". *Studies in History and Philosophy of Science* A, 31 (1), 119-37.

Glas, Eduard (2001b). "The 'Popperian Programme' and Mathematics. Part II: From Quasi-Empiricism to Mathematical Research Programmes". *Studies in History and Philosophy of Science* A, 32 (2), 355-76.

Goudge, Thomas A. (1969). *The Thought of C.S. Peirce*. Dover Publications, Mineola, NY (Originally published by the University of Toronto Press in 1950.)

Hallett, Michael (1979). "Towards a Theory of Mathematical Research Programmes I". *The British Journal for the Philosophy of Science*, 30 (1), 1-25.

Hallett, Michael (1979). "Towards a Theory of Mathematical Research Programmes II". *The British Journal for the Philosophy of Science*, 30 (2), 135-59.

Kadvany, John (1995). "The Mathematical Present as History". *Philosophical Forum*, 26 (4), 263-87.

Kiss, Olga (2002). "Mathematical Heuristic - Lakatos and Pólya". In George Kampis, Ladislav Kvasz, and Michael Stöltzner (Eds.) *Appraising Lakatos*. Kluwer Academic Publishers, Dordrecht, 243-54.

Kitcher, Philip (1983). *The Nature of Mathematical Knowledge*. Oxford University Press, New York.

Kitcher, Philip (1995). "Arithmetic for the Millian". In Michael Resnik (Ed.) *Mathematical Objects and Mathematical Knowledge*, Dartmouth, Aldershot, 155-76. (Reprint of the 1980 article published in Philosophical Studies 37: 215–36.)

Koetsier, Teun (1991). *Lakatos' Philosophy of Mathematics. A Historical Approach*. North-Holland, Amsterdam - London - New York - Tokyo.

Koetsier, Teun (2002). "Lakatos' Mitigated Scepticism in the Philosophy of Mathematics". In George KAMPIS, Ladislav Kvasz, and Michael Stöltzner (Eds.) *Appraising Lakatos*. Kluwer Academic Publishers, Dordrecht, 189-210.

Kvasz, Ladislav (2002). "Lakatos' Methodology Between Logic and Dialectic". In George KAMPIS, Ladislav Kvasz, and Michael Stöltzner (Eds.) *Appraising Lakatos.* Kluwer Academic Publishers, Dordrecht, 211-41.

Lakatos, Imre (1972). "Falsification and the Methodology of Scientific Research Programmes". In Imre Lakatos and Alan Musgrave (Eds.) *Criticism and the Growth of Knowledge.* Cambridge University Press, London - New York, 91-196.

Lakatos, Imre (1976). *Proofs and Refutations. The Logic of Mathematical Discovery.* Cambridge University Press, Cambridge (Edited by John Worrall and Elie Zahar.)

Lakatos, Imre (1998). "A Renaissance of Empiricism in the Recent Philosophy of Mathematics". In Thomas Tymoczko (Ed.) *New Directions in the Philosophy of Mathematics. An Anthology.* Princeton University Press, Princeton, NJ, 29-48. (First published in his Mathematics, Science and Epistemology. Philosophical Papers Volume II in 1978, posthumously edited by John Worrall and Gregory Currie, 1978, 24-42.)

Larvor, Brendan P. (1998). *Lakatos: An Introduction.* Routledge, London.

Larvor, Brendan P. (2001). "What is Dialectical Philosophy of Mathematics?" *Philosophia Mathematica* (III), 9 (2), 212-29.

Laudan, Larry (1977). *Progress and Its Problems.* Routledge and Kegan Paul, London.

Laudan, Larry (1987). "Progress or Rationality? The Prospects for Normative Naturalism". *American Philosophical Quarterly*, 24(1), 19-31.

Losee, John (2001). *A Historical Introduction to the Philosophy of Science.* Oxford University Press, Oxford (Fourth edition of the 1972 original.)

Losee, John (2004). *Theories of Scientific Progress. An Introduction.* Routledge, New York - London.

Maddy, Penelope (1997). *Naturalism in Mathematics.* Clarendon, Oxford.

Mancosu, Paolo (2000). "On Mathematical Explanation" In Emily Grosholz and Herbert Breger (Eds.) *The Growth of Mathematical Knowledge*, Kluwer Academic Publishers, Dordrecht - Boston - London, 103-19.

Milne, Peter (1994). "The Physicalization of Mathematics". *The British Journal for the Philosophy of Science*, 45 (1), 305-40.

Murawski, Roman (1999). "On New Trends in the Philosophy of Mathematics". In Ewa Orlowska (Ed.), *Logic at Work.* Dedicated to the Memory of Helena Rasiowa, Springer-Verlag, Heidelberg - New York, 15-24.

Oliveri, Gianluigi (1997). "Criticism and Growth of Mathematical Knowledge". *Philosophia Mathematica* (III), 5 (3), 228-49.

Pólya, George (1962). *Mathematical Discovery. On Understanding, Learning, and Teaching Problem Solving. Volume I.* John Wiley and Sons, New York.

Pólya, George (1965). *Mathematical Discovery. On Understanding, Learning, and Teaching Problem Solving. Volume II.* John Wiley and Sons, New York.

Pólya, George (1957). *How to Solve It.* Princeton University Press, Princeton, NJ (Second edition of the 1945 original.)

Pólya, George (1968). *Mathematics and Plausible Reasoning. Volume II*: Patterns of Plausible Inference. Princeton University Press, Princeton, NJ. (Second edition of the 1954 original.)

Pólya, George (1973). *Mathematics and Plausible Reasoning. Volume I: Induction and Analogy in Mathematics.* Princeton University Press, Princeton, NJ (Eight print of the 1954 original.)

Popper, Karl R. (1978). *Conjectures and Refutations. The Growth of Scientific Knowledge.* Routledge and Kegan Paul, London - Henley (Reprint of the 1972 fourth revised edition of the 1963 original.)

Popper, Karl R. (2002). *The Logic of Scientific Discovery*. Routledge, London - New York (Reprint of the 1992 edition of the 1959 translation published by Hutchinson and Co, originally published in German in 1934.)

Resnik, Michael (1980). *Frege and the Philosophy of Mathematics*. Cornell University Press, Ithaca, NY.

Sandborg, David (1998). "Mathematical Explanation and the Theory of Why- Questions". *The British Journal for the Philosophy of Science*, 49 (4): 603-24.

Schoenfeld, Alan H. (2000). George Pólya and Mathematics Education. In Gerald L. Alexanderson (Ed.), *The Random Walks of George Pólya*, The Mathematical Association of America, Washington, DC, 233-6 (Reprint with postscript of the 1987 original published in Bulletin of the London Mathematical Society 19.)

Shapiro, Stewart (2000). *Thinking about Mathematics. The Philosophy of Mathematics*. Oxford University Press, Oxford.

Steiner, Mark (1983). "The Philosophy of Mathematics of Imre Lakatos". *Journal of Philosophy*, 80 (9), 502-21.

Tiles, Mary (1996). "Philosophy of Mathematics". In Nicholas Bunnin and E.P. Tsui-James (Eds.) *The Blackwel l Companion to Philosophy*, Blackwell, Oxford - Cambridge, MA, 325-57.

Torretti, Roberto (1984). *Philosophy of Geometry from Riemann to Poincaré*. D. Reidel Publishing Company, Dordrecht - Boston - Lancaster (Paperback reprint of the 1978 original edition.)

Tymoczko, Thomas (Ed.) (1998). *New Directions in the Philosophy of Mathematics. An Anthology*. Princeton University Press, Princeton, NJ (Second, revised and expanded edition of the 1985 original published by Birkhäuser.)

Van Bendegem, Jean Paul (1990). "Review Article: Scarre Geoffrey, Logic and Reality in the Philosophy of John Stuart Mill". *De Uil van Minerva*, 6 (3), 202-4.

Yuxin, Zheng (1990). "From the Logic of Mathematical Discovery to the Methodology of Scientific Research Programmes". *The British Journal for the Philosophy of Science*, 41 (3), 377-99.

INTERLUDE 9

There was a time—although in private we think this time has never really ceased—when the question “Is God a mathematician?” was considered a serious conundrum whose answer would reveal deep truths about the world, ourselves, and God. Bart Van Kerkhove already made clear in his paper that in our mathematics we need not necessarily reserve a place for God; or, to quote a famous mathematician, “If God has mathematics of His own, let Him do it Himself.” Well, at least not in the mathematics we actually practice, as he tried to show.

Did you notice the curious term we used in the paragraph above? We spoke about *our* mathematics. Does that make sense? To paraphrase Wittgenstein, what can you possibly mean if you are saying that “2 + 2 = 4 in Ghent on Friday the 22nd of July, 2007 at 8:00 pm?” But we do know that different cultures have different mathematics. So it does make sense to speak about *ours* and *theirs*. Unfortunately to many it also makes sense to conclude, on the premise that there can be only one mathematics, that there should really be an orderly relation among the different kinds of mathematical systems. Or, to put it bluntly and in semi-formal language to increase its credibility: *ours* > *theirs*.

There is an alternative way to formulate what we are claiming here. The question “Is God a mathematician?” should really read “Is God a Mathematician?” In other words, using the well known device of Alan Bishop, perhaps it is all right that “God is a Mathematician” with a capital *M*, but does it make sense to ask “Is God a mathematician?” with a small *m*. Is it not much more meaningful to talk about all the diverse forms of mathematics with small *m* and to reserve in a very specific *niche* of its own, *our* Mathematics with capital *M*? Therefore, we have no orderly relation, but rather a network of mutually inspiring systems and/or practices of mathematics, each one forcing us to reflect about what a mathematics education could be like, should be like, but usually isn’t. Nevertheless, the contribution of Rik Pinxten and Karen François shows us that it is not an impossible task to transform, at least for once, *could* and *should* into *is*.

ETHNOMATHEMATICS IN PRACTICE

Rik Pinxten and Karen François
Ghent University, Belgium - Free University of Brussels, Belgium

Abstract: In this paper we elaborate on the difference Bishop made between mathematics with a capital *M* and mathematics with a small m. The relation between *M* and *m* is far from clear and, confronted with the task of teaching mathematics, the problem only sharpens. In the first section (Ethnomathematics) we give a brief presentation of the critical role of mathematics. We make clear that we should be conscious of the institutional aspects of mathematical learning and teaching. Mathematics education always takes place in an institutional context, e.g. the school context. The point is that mathematical learning or thinking is contextual in any living culture; it lives and develops and is used in a particular cultural context. We make a plea to consciously and explicitly seize and actively practice these different world views, to have pupils attain a level of comprehension and sophistication. This practice should be an ingrained part, implicitly or unconsciously, of the curriculum of mathematics. In the second section (Empirical Facts) we offer some suggestions about the practical use of ethnomathematics in the classroom, taking as a starting point examples of field research among the Navajo Indians (in the U.S.) and among the Turkish ethnic minority (in Belgium).

Key words: Ethnomathematics, Navajo Indians, Turkish ethnic minority, *Hooghan*

1. INTRODUCTION

Bishop (1988a) rightly stressed the difference between *M*(athematics) and *m*(athematics). *M* is mathematics as it is known and developed as a scientific discipline, for example at universities. In other words, it is the mathematical knowledge of the mathematician and of the highly specialised student in physics or engineering. Opposed to this is *m* as the set of skills and procedures to count, to measure, and the like, which an individual uses in

daily life. There is probably a certain degree of overlap between them, but on the whole the relation between *M* and *m* is far from clear. In mathematics education this problem is posed even more poignantly: should we accept the idea that the mathematician's *M* ought to be learned by everyone, or should we develop the *m* in the subjects' culture through our mathematics classes? Are these choices exclusive, as is often said? We will offer some suggestions about this matter, taking as a starting point examples of field research among the Navajo Indians in the U.S. and among the Turkish ethnic minority in Belgium.

2. ETHNOMATHEMATICS

Partly because of the growing crisis in mathematics education and partly because of fundamental problems in the so-called *absolutist* views in the foundations of mathematics, a naturalist theory of mathematical knowledge is currently gaining the field. The *absolutist* views regarding mathematics (formalism, logicism, and constructivism; see Ernest 1991) claim that *M* knowledge is *a priori* and indisputable. They believe that mathematically true statements are always logically deducted from *a priori* or given propositions. According to this school, empirical reality does not intervene. Consequently, mathematics education had better follow the structure of *M* and initiate pupils in the deductive and thus fundamentally decontextualized (or not reality-bound) knowledge of the mathematical field. Ideally, one can start by teaching the concepts and the procedures belonging to the *basement* of the mathematical building (e.g., set theory), and logically continue climbing up to every next *floor* (e.g., arithmetic, algebra, etc.). The philosophy of the so-called "Modern Mathematics" (in programmed form or otherwise) is the most radical version of this view. The concepts and procedures of *m* are then regarded as "bad mathematics", false procedures, or at best pre-scientific versions of *M*. They need to be stamped out or thoroughly corrected by education.

An alternative view of *M* is now gradually developing. The failure of the absolutist foundation approaches (shown by Gödel, but also proved by the contradictions within absolutist visions: see Ernest 1991) and the increasing collapse of mathematics education programs have given rise to more naturalistic theories about mathematical knowledge. One theory argues that mathematics can be seen as *essentially a symbolical technology* (Bishop 1988a, 18), based on skills or environmental activities of a cultural nature. Another theory is rightly labelled *ethnomathematics,* as we see it. In general, proponents of the naturalistic approach claim that mathematical knowledge (including insights and intuitions of *M*) is rooted in the cultural context of

the knower. Knowledge in general—and mathematical knowledge in particular—is cultural and contextual by nature (Pinxten 1992). One consequence of this position is that *M*, in a way that is not trivial, builds on and is led by concepts and intuitions of *m*: empirical facts, arguments and even action procedures of someone's culture provide the real contextual foundation for *M*. The rest of this article will focus on this perspective of mathematics only.

One question which automatically crops up in relation to such an approach to mathematics is: what is *m* like? It is this selfsame question that the chief part of ethnomathematics seeks to answer. In our world, in all cultures that we know of, people develop activities which lay the foundations for formal or mathematical knowledge. Bishop (1988a and 1988b) distinguishes six different types of activities: counting, locating in space, measuring, drawing, playing, and explaining. Different cultures engage in and develop these activities in different ways, and the study of particular forms in each culture will yield an insight into mathematical skills and the ways that they are exercised and learned in those cultures.

Over the years various researchers have discovered and documented one or more of these activities in a multitude of cultures. Some early works include, of course Zaslavsky's (1973) about counting in Africa, and Needham's monumental volume about algebra and geomancy in Old China (1965). Lancy (1983) offers interesting information about Papua skills and traditions, and I myself (Pinxten) have mainly concentrated on spatial knowledge and theory of forms among Navajo Indians (Pinxten et al. 1983 and 1987).

The first comprehensive work in this field is the outstanding book by M. Ascher (1991). Ascher is obviously a mathematician and not a social scientist (like Lancy and me). Her approach therefore starts from a deep insight into *M*. Being rooted in this disciplinary background, she selects the concepts and skills in our mathematical experience which seem most relevant and interprets other cultures' varied activities and knowledge in this light. In this manner, numerical understanding (and not just counting as an activity), for example, is studied as an implicit or explicit concept in various cultural phenomena. Ascher develops an deep-structural analysis, beyond the mere description of terminology and actions which numbers express in a cultural community, and thereby shows how numbers operate in observable cultural phenomena. The same goes for graphs: complex drawings of so-called geometrical figures in the sand are systematically analyzed as an example of the use of graphs; spatial conduct, geometrical decorations, the logic of family relations and diverse other cultural phenomena are thus described and explained on the basis of mathematical concepts and procedures that are used or expressed through them. In our opinion, the

importance of Ascher's work is twofold. In the first place, she develops a complement to the social-scientific studies in this domain: instead of describing phenomena in terms of the subjects' culture or psychology, she takes the other road and describes the cultural and psychological givens as ways to tackle the mathematical problems which she recognizes. This adds another dimension to the *soft* approach of anthropologists and psychologists. In the second place, we recognize a solid bridge between M and m in her work. Indeed, where social scientists have given detailed descriptions of at least part of m (i.e., the mathematical knowledge of the layperson in a particular culture), Ascher provides a reinterpretation of these facts in terms of M (i.e., the concepts of mathematics as a discipline) and demonstrates how m and M can be linked to one another. This allows for the development of an explicit relation. In other words, an educational program can be set up which would lead learners from their m to M. This perspective takes D'Ambrosio's (1987) broad aim one step further: the scheme can finally be implemented. It is with regard to this that this current work wishes to make a small modification.

3. EMPIRICAL FACTS: NAVAJO INDIANS AND THE TURKISH ETHNIC MINORITY

In 1976-1977 and 1981, fieldwork was undertaken (together with Ingrid van Dooren: Pinxten et al. 1987) which led to a curriculum booklet for geometry teaching among Navajo Indians. Facts about the Turkish ethnic minority and their spatial knowledge were mainly collected by Marijke Huvenne, an assistant for a project of immigrant studies in our department (University of Ghent).

3.1 Navajo Material

During our research on Navajo Indians' spatial knowledge, various things struck us. Navajo experience their world as a fundamentally dynamic world: nothing is still, in fact, there is only movement and change. Their world consists of events and processes, whereas ours consists first and foremost of situations and objects.

There is a correlate in the language: Navajo language expresses almost everything in verb forms, whereas Indo-European languages make a distinction between (pro)nouns and verbs. In Navajo the verb *to go* is conjugated in no less than 300,000 ways, yet there is no counterpart for *to be*. Many aspects of the language describe a myriad of ways in which actions begin, stop, are repeated, etc. As a consequence of this manner of speaking

and thinking Navajo do not use a part-whole logic in their world view. In Western views this logic is the basis of our formal (and daily) thinking: we see the world of experience as a whole which can be split up in many parts. We thus tackle a problem by splitting it into parts that are easier to deal with. The sum of all partial solutions then leads us to the answer to the whole problem. In mathematics education, too, this part-whole characteristic of our world view is very prominent: set theory distinguishes between the set (the whole) and its elements (parts), with special operations defined for both; in geometry, a line is the set of all points; etc. In the Navajo world view such part-whole reasoning is non-existent.

A second characteristic is that Navajo concepts of space and forms are almost always determined by movements or shifts: a limit or side is not a line, but an obstacle which modifies a running act (like stopping, transgressing, etc.); distances are thought of in terms of running or moving and of specific characteristics of landscape. In the curriculum booklet on theory of forms which we have developed, these specific emphases are incorporated thoroughly. As we take the naturalist view of mathematical knowledge to be the most sensible, the booklet explicitly and consciously starts with the spatial "world as it is experienced by the Navajo children." So we selected different contexts of experience in the children's preschool cognitive world and worked with these to spur them on toward a more sophisticated geometrical cognitive development. The difference in detail between this and existing ethnomathematical propositions is that we propose to take the autochthonous categories seriously and to develop, practices, and refine them as much as possible. This could take many years of training, wherein only the Navajo language and Navajo concepts of space and form are dealt with. Only after this period, and on the basis of the insights which have by then become conscious in one's own spatial perspective, can the step towards Euclidean, or *Western*, geometry and mathematics be taken. Without this, mathematics education for most children and adults in this culture remains abortive; the necessary leaps of insight have not been made, because of the blindness to Navajo interpretations of space and world throughout the teaching process. Indeed, the mathematics program as it is normally taught is Western-based—i.e., it starts from a Western world view that remains implicit and that, taking its *scientific* basis for granted, is thought to be typical of all people and cultures.

The mathematics teacher ought to become conscious of the cultural *baggage* or bias in his/her curriculum. Visualization is a crucial element in this teaching process, from exploring one's own categories to the conscious exploration and sophistication of geometry and proper mathematics. The relation between m and M is not automatic in my view, and should first be developed in the autochthonous form. An example will clarify this.

One of the contexts of experience which we have elaborated is the *hooghan* (traditional housing). The *hooghan* is a cosmological scale model of the world and illustrates the Navajo concepts for above-below, proportions, wind directions, room to move, etc. The child knows all this by visiting *hooghans*, by living in one, or by receiving explicit explanations about it. The *hooghan* is also a didactic instrument. In the classroom children are invited to mentally explore all aspects of the *hooghan* and to reconstruct a *hooghan*. A group of children work in a team to build a *hooghan* with refuse. During this process, notions of orientation and proportion are explored. In the second phase, a scale model is built. The visualization process becomes more intense and the group must agree about size, proportions, and some technical terms. Finally, graphic presentations of the *hooghan* are made, leading to a more abstract conception. Navajo language is used all the time and autochthonous concepts are respected and developed further by these lessons.

In the following section, we will elaborate on the example of the *hooghan* project, which is only one example among others (e.g., rodeo, school compound, herding sheep, and rug weaving; Pinxten et al. 1987, 63-69).

3.1.1 Project: the *Hooghan*

The *hooghan* is the single most important construction of the Navajo tradition. Although more and more Navajos go to live in new prefabricated houses or trailers nowadays, the *hooghan* is not only still in use by many, but it remains the necessary ceremonial construction for all. Thus, it is not rare to see people live in a house or trailer and build a *hooghan* on the side for ceremonial purposes. Again, although the rather primitive and barely furnished traditional form of the *hooghan* might be vanishing rapidly now, the modern and often well-equipped *hooghans* still have the same basic form as the traditional one (McAllester 1980).

It is to be expected that all children know *hooghans* from their personal experience, either because they assisted a ceremony or because they have lived in one for a longer or shorter period (with grandparents, for example). Not only that, but because of the fundamental structural characteristics of the building, all children will have some knowledge of the meaning and the form of the *hooghan*.

3.1.2 Constructing a *Hooghan* Scale Model

Children will go and visit a *hooghan* in the neighbourhood. They will ask about the way it was constructed, and eventually assist at the construction of

a new *hooghan* in the vicinity. They will note down what they learn and explain everything in the classroom. Beginning with this material, they proceed to construct a scale *hooghan* in the classroom, again using simple materials and mud most of all.

Apart from this, we advise simulating a *hooghan space* in the classroom, e.g., by sitting together in the proper way, surrounded by a "wall" of carton boxes with only one opening to the east. With this preparation, the exploration period can start. In the construction of a scale model *hooghan* or of a *hooghan space* to sit in, the following elements of the sturcture should be kept in mind whatever variation of form one might use: the *hooghan* has a center, it has one door opening to the east, and it is more or less circular. A first rough approximation of the *hooghan* can be simulated; during the lessons the faults and difficulties of this construction will be detected, after which a more accurate version can be tried out.

3.1.3 Exploration

"Playing *hooghan*": The children are visiting their grandparents who live in a *hooghan.* What should they do? When is eating time, and where will the meal be cooked? When is sleeping time, and where do they sleep? From examples and the talk of the children, it will appear that the center is the main structural element of the internal space of the *hooghan*: people sleep around the center, walk around it, etc. The center is where the fire usually is and where the sand painting ceremony will be held. Around the center one can detect the walls and the poles in all cardinal directions. The notion of center can now be explored more systematically by the following exercises:

- Try to point to the center of the *sit-in-hooghan* in the classroom.
- Walk from any "wall" towards the center and count your steps.
- Adjust the center of the *hooghan* until it is more accurately in the middle of the *hooghan* space.
- How can we know the distance between center and "wall"? —by stepping. But the length of steps differs with the length of one's legs. You can explore this with teachers and students stepping. How can we get to a uniform measure? —by using rope or stick. One chooses a rope or stick and reconstructs the distance between center and "walls" in a more accurate way. To practice the notion of *measure*, smaller and bigger *hooghans* can be constructed in order to grasp the idea of interdependence between center and distance (width) by using different lengths of sticks or ropes. One should always keep each of the ropes or sticks used, in order to aid visualisation.

The structure of the *hooghan:* A *hooghan* is a bowl-shaped form, consisting of a shell-like top and a shell-like bottom. These represent heaven and earth in Navajo tradition. Those forms are constructed by defining a center and four cardinal directions (east, west, south, and north). These can be determined in the classroom and oriented the proper way. Around the four main poles (one in each direction) sticking straight up in the sky, legs or planks are attached, one on the top on the other. In this instance the difference between vertical and horizontal extension is clear. The horizontal logs reach a certain height (man-high); on top of this construction the roof is built. So what you have in different stages of construction is the following:

- A flat surface with center and peripheral walls.
- A determination of the border of the *hooghan* by indicating the place of the vertical poles and the horizontal walls.
- This gives a sort of open cylinder, which is then topped off by a roof.

The children will try to represent this process in drawing and in actual manufacture. Once the principles of the construction procedure are more or less grasped, one can go on and represent the model of the *hooghan* at a more abstract level by using folding figures (in the sense of the UNICEF Educoll system).

Figure 1. Representing the model of the *hooghan* at a more abstract level.

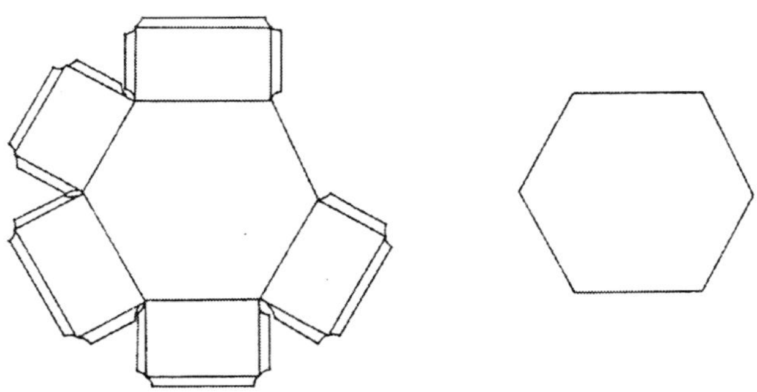

In Figure 1, above, the *hooghan* has six sides, one of them being the door opening. The figure should be prepared by the teacher and can be cut and colored by the children. The lid or roof is somewhat problematic. It might be best to use a flat hexagon and fill it up if need be with some waste material to have it bulb out a little bit.

One can now explore the *hooghan* shape in a systematic way. While turning the *hooghan* model around, it looks approximately the same from all directions. This corresponds with the real *hooghan*: its geometric shape (hexagon or octagon) makes it appear more or less identical from any angle. The following exercises in visualization can now be performed:

- Looking at it from one direction, one sees a rib extending towards one's eyes in front, whereas the back ribs cannot be seen. Is this rib bigger than the other ones? Why do we see it as bigger and not see the other ones?
- Turning the paper model around, we always have a similar appearance (except for the door) however we turn it. How is this possible? In this way the uniformity of the geometric shape is perceived and consciously understood.
- Looking at the top of the *hooghan* (as from the rim of a canyon down onto the *hooghan* in the canyon), we perceive the polygon as a whole. When looking directly from above, we cannot see the entrance, and the polygon appears to be regular on all sides, with a hole in the center (the smoke hole).

Once this point is grasped, children will try to draw the *hooghan* from different perspectives. These drawings will be imprecise, because the children cannot master the use of the drawing instruments of geometry. Nevertheless, they will try to approximate the form in a graphic representation. They will gradually learn that they should start from the center position and measure a uniform rod distance from there. Eventually, a stick or a piece of rope can be used to contruct the polygon, as shown in Figure 2.

Figure 2. Constructing the polygon.

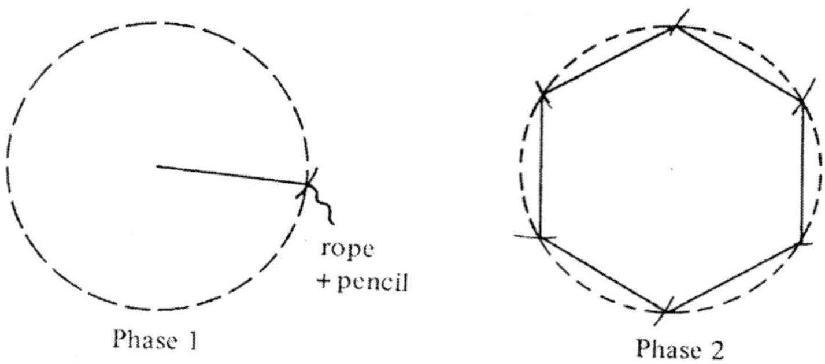

By means of these drawing exercises, it will become clear that the polygons fit nicely into a circle and that many polygons can be fitted into the same circle. This may lead to the drawing of different maps of *hooghans*: *hooghans* with six, eight, or ten walls, and perfectly round *hooghans* (like the mud *hooghans* that can be seen in some places still).

Details: The logs or planks which are used to erect the walls of the *hooghan* have a particular shape. By means of interviews the children will become aware of this shape and the functionality of it in the construction: What shape do they have? (straight, smooth, just long enough for one side). Why would that be so? (it is easier to build with flat and straight logs or planks than with irregular or bulky ones; they fit nicely on top of each other).

By imitating the construction of *hooghan* walls by means of building block games, it will become clear that straightness along the length dimension and regular straight angles for the thickness dimension make it easier to put one on the other. By making corners (to reach the hexagon figure), the construction gets more stability: each wall supports the other ones. This can be easily tested by the children: construct a wall of light material and try to push it. The wall easily gives way. Now do the same to make a cube or a hexagon: the construction is much more stable and cannot be pushed so easily. Even human beings making such forms by their intertwined bodies can illustrate this principle (children in classroom).

This exercise teaches about the use of flat surfaces (walls) to make a stable space (*hooghans*). During this process, we take up notions like straight, angle, surface, hexagon, bottom/top, and center.

The above mentioned aspects of space can be handled abundantly in drawing lessons.

- Make a plan of a *hooghan* with eight sides.
- Draw the walls next to one other as flat surfaces.
- Develop different perspectives on the *hooghan* and represent these simply in drawings:

  a) Looking from the top down (from the rim of a canyon), we see the polygon structure with center and walls;
  b) Looking at it in a horizontal plane (as when coming up to a *hooghan* when sheep herding, looking at it up front), we see a rectangle with a bulbing roof, as in Figure 3.

Figure 3. Looking at a *hooghan* in a horizontal plane.

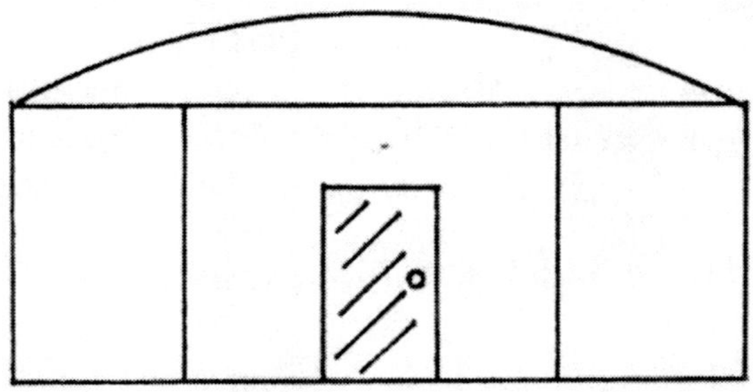

c) Looking at the dirt *hooghan* from the side, we see a half sphere. This shape can be studied some more. First it should be drawn, as in Figure 4. What are the characteristics of this shape? Use the rope to detect these characteristics (from center out, the whole circumference can be drawn, ending in a straight line on the surface). Suppose we put two dirt *hooghans* on top of each other, one upside down, and touching each other's base, what would we have? (A sphere). What can you do with a sphere? (Roll it on every side). When is it standing up? When is it lying sideways? Children will then look for other spheres in their environments (like balls, tumble weeds, etc.). They will explore the peculiarities of the sphere and try to reconstruct one in plasticine or other material.

Figure 4. Looking at a dirt *hooghan* from the side.

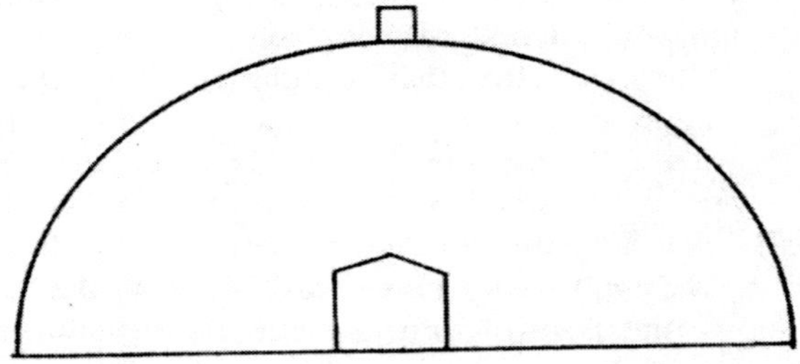

Conclusion: Children should gradually grasp the terminology (standardized for the purpose), the graphical representation, the situational and the process aspects of all geometric figures in this context. We think it is clear that one should always start from the natural context, then visualize the features in it and turn to graphical representation. This holds true in the *hooghan* project as in the other projects. The *hooghan* project is one of the five projects used in Navajo geometry teaching. In addition to these, we started similar projects with children of the Turkish minority in Flanders (Belgium).

3.2 The Turkish Ethnic Minority

The Turkish ethnic minority in Belgium comes predominantly from rural areas in Turkey. Their language and world view can to a considerable degree be called pre-Newtonian, although their distance to European subjects is of course much smaller than that between Navajo and European worlds of experience. During our observations we particularly found two differences in the daily world of experience. The tasks in mathematics education are often ethnocentric (e.g. about Western city life instead of about Turkish country life), which for Turkish pupils causes misunderstandings. Moreover, the Turkish language is sometimes a source of difficulties in a Western school context: plurals are not shown in nouns, but an appellative usually replaces numerals and the plural. In Turkish, there is no such thing as a prefix, which leads to difficulties with *self-evident* spatial references like above, under, beside, etc. These few remarks give an idea of the problem confronting us. We propose the following curriculum project to overcome the difficulties.

Children in mixed class groups (migrants and autochthonous) are asked to discuss their neighborhood. Concepts like center, border, near and far off, etc. are described in both Turkish and Dutch. The children are encouraged to work together across cultural borders. Translation problems that crop up in the process are largely left to the children themselves. The building of scale models and the development of graphic presentations of the neighborhood allow them to get a grip on the differences and communalities in meaning and speech about abstract notions such as distance, size, proportion, geometrical shapes (after all, we are working in a highly geometricised urban environment), etc. It is only after this initial period of consciousness-building and training in visualising that the leap towards Euclidean geometry and mathematics in the strict sense may be given a try. In this way, insights into the functionality and the use of geometrical notions and into the relevance of precise agreements will become clear to the children. They can thus lay the foundations for a better take-off with regard to proper

mathematics education. The mixed class groups stimulate a mutual understanding and a deeper awareness of the difficulties and the possibilities of a multicultural learning environment.

At the same time, it seems to be important to be conscious of the institutional aspects of mathematical learning. Mathematics education always takes place in an institutional context. The context that is dominant, and for which both teacher training and curriculum development programs have been set up over the years, is the school context. The point seems to be that mathematical learning or thinking is contextual in any living culture; it lives and develops and is used in particular cultural contexts. The often decontextualized use of knowledge in the Western school setting is then alienat and foreign to the pupils. Our experience with pupils of the Turkish minority gave us the following impression: for them the school is felt to be an alienating but necessary instrument to reach a better position in life. In their case the heavy emphasis on rote learning, which at least correlates with the Koran school paradigm of learning, appears to me as a compensation for the lack of insightful learning. At the same time, the present understanding in Belgium about the lag in school results between native (Flemish) and non-native (Turkish) children is explained also by the different familiarity in each group with the institution of the school. For example, we find on our research that the children of the Turkish minority continue to have duties at home (e.g., washing dishes, caring for newborns, etc.) after they finish school, forcing them to postpone homework until late in the evening. Their parents are either unable or unwilling to help them with tasks for school because the parents are illiterate in the Flemish language or because they say, "That is the task of the school." The parents motivate their children to perform well in language but much less or not at all in mathematics classes. These and similar findings seem to suggest strongly that school as an institution must be felt as doubly alien by immigrant children: it has little or nothing to do with life in their *home culture* and it is a foreign world they have to confront all by themselves. Motivation turns out to be problematic; and special programs, geared to overcome these problems of retardation in school, are hence often seen as indicating and continuing deficiency rather than as helping with the learning difficulties of the children.

4. CONCLUSION

Even this very brief presentation about these cases may have clarified to the reader what it is that we wish to stress in this article. In general, we agree with the ethnomathematics approach as we understand it. There is just one point which we feel is lacking and which we want to mention here. We want

to suggest, on the basis of this action research, that children of different cultures live with another world view than those implied in the mathematics curricula that are used in education. Our plea is to consciously and explicitly seize and actively practice this other world view, to have pupils attain a level of comprehension and sophistication which forms an ingrained part, either implicitly or unconsciously, of the curriculum of mathematics.

Exploring the functionality relations and spatial concepts in the children's mother tongue and in their real worlds of experience can bring this about.

Through such exploration, linguistic processing, and the true exercise of visualizing, the child will attain the insights needed to grasp the underlying (implicit) world view of M. That is why the transfer from m to M, and the use of m concepts when developing M successfully, will need to take place via the conscious, systematic, and explicit exploration of the largely unconscious and half-developed perspective of m which preschool children of another culture possess. The choice between opting for a version of M which is seen as standard, or leaving room for alternative elaborations of mathematical thinking, is not dealt with here. A conservative and an unorthodox attitude remain possible.

REFERENCES

D'Ambrosio, Ubiratan (1987). *Socio-cultural Bases for Mathematics Education*, Unicamp, Campinas, Brasil.

Ascher, Marcia (1991). *Ethnomathematics, A Multicultural View of Mathematical Ideas*, Brooks/Cole, Pacific Grove-California.

Bishop, Alan. J. (1988a). *Mathematical Enculturation, A Cultural Perspective on Mathematics Education*, Kluwer Academic Publishers, Dordrecht.

Bishop, Alan J. (1988b). The Interactions of Mathematics Education with Culture, in: *Cultural Dynamics, An International Journal for the Study of Processes and Temporality of Culture*, Volume I. 2, 145-157.

Ernest, Paul (1991). *The Philosophy of Mathematics Education, Studies in Mathematics Education*, RoutledgeFalmer, London.

Lancy, David F. (1983). *Cross-cultural Studies in Cognition and Mathematics*, Academic Press, New York.

McAllester, David P. (1980). *Hogans: Navajo houses and house songs / house poems from Navajo ritual.* Photos of Navajo houses by Susan W. McAllester, Columbia University Press, Middletown, Conn.: Wesleyan University Press ; Irvington, New York.

Needham, Joseph (1954) (1965). *Science and Civilization in China*, Cambridge University Press, vol. 3, Cambridge.

Pinxten, Rik; Van Dooren, Ingrid and Harvey, Frank (1983). *Anthropology of Space, Explorations into the Natural Philosophy and Semantics of Navajo Indians*, University of Pennsylvania Press, Philadelphia.

Pinxten, Rik; Van Dooren, Ingrid and Soberon, Erik (1987). *Towards a Navajo Indian Geometry*, K.K.I. Books, Gent.

Pinxten, Rik (1992). Knowledge as a Cultural Phenomenon, in: *Cultural Dynamics, An International Journal for the Study of Processes and Temporality of Culture*, Volume V, 83-104.

Zaslavsky, Claudia (1973). *Africa Counts*, Prindle, Weber and Schmidt, Boston.

POSTLUDE

Alan J. Bishop

When Karen François and Jean Paul Van Bendegem invited me to write a postlude to this volume, it was with a slight sense of trepidation that I agreed. My trepidation arose from two quarters—firstly, I am not a philosopher, and second, at that stage I had not seen many of the chapters of the book.

I also should express one other issue that I considered—that is, as the editor for this book series, I have always been reluctant to give an impression that it should contain only those ideas and interests which I personally wish to promote. Indeed the reason for having an impressive and authoritative editorial board for the MEL series is precisely to counteract and moderate any bias which I might have in facilitating the books which are offered to the series. This is particularly the case with this book, for as the reader will have seen, some of the developments in the book specifically refer to the main ideas of my 1988 book *Mathematical Enculturation*. That, of course, was one of the reasons that the editors of this book invited me to contribute this brief postlude.

My pursuit of the consequences of adopting a cultural perspective on mathematics education led me first into the ideas of ethnomathematics, cultural diversity, and the negative social consequences of continuing to adopt a non-cultural, or *universalist* approach to mathematics education. Fortunately this field has developed hugely since the 80's, and as a result of many people's efforts it would be surprising if a young mathematics education researcher today would be ignorant of that field, and of its implications for educational research and practices. As we have seen in the chapters of François, and Pinxten and François, the ideas of ethnomathematics still challenge the assumptions and philosophy underlying

the so-called universalist mathematics education still practiced today in most countries.

I was also delighted to see many of the authors' interest in the second of my consequential explorations from the cultural perspective—namely, the role and development of a mathematics education which recognises the existence of values and valuing. If mathematics is a form of cultural knowledge, I argued, then it must also involve certain values. Consequently, mathematics education must also learn how to deal with those values. The core theme of this book, as stated by the editors in the prelude is "Making the implicit explicit and asking questions" It is a theme that I am happy to share—particularly as through my research on values it is clear that there is much to be done to make sure that the many *implicits* and *assumptions* in mathematics education become more explicit and questioned.

Much of my research interest in values relates to the fact that one can often find references in mathematics curriculum documents to values, but there is never any follow-up as to how any of the values might be addressed in the mathematics education which is being encouraged or prescribed. Thus there is a challenge for us in research. If values are a significant feature of mathematics education, why is there so little known, written, and practiced about them? Why are values so implicit in mathematics education?

I believe the main reason concerns the relationship between mathematics and science, and in particular the shared belief of Western Mathematics and Science (henceforth capitalized) in theoretical universalism – the idea that Mathematics and Science knowledge is universal and in that sense non-culturally specific. If there is no reference to culture, there is likely to be no reference to values. The fact that all of that is implicit does *not* mean that universalism is not valued. Of course it is; and it is practiced too, in implicit and covert ways.

Ethnomathematics represents a huge challenge to the Mathematics community precisely because it challenges their fundamental belief in, and assumption of, universalism. Nowadays, however, an equally provoking challenge is coming from the idea of *numeracy*. This construct has many meanings. Some see it as *practical math*, some as *applied math*, some as *everyday math*, and some as *out-of-school math*. Some see it as involving only numerical calculations, while others see it as the whole of school mathematics.

I would like to use the context of this book to raise another possibility. I see the relationship between numeracy and Mathematics being based on the idea of Mathematics as theory, with its chief educational role being to explain, and to extend, numeracies and ethnomathematical practices.

Mathematics intersects with the scientists' *world* also through the use of mathematical expressions, formulas, and concepts. Indeed modern science

progresses largely through the use of both traditional and modern mathematical constructs. Mathematics, however, can also be used to explain numeracy and ethnomathematics practices, and here are a few simple examples:

- Why do the different algorithms for addition, subtraction, multiplication, and division work, and give the same results?
- Why can one calculate in different bases?
- Why do planes fly the great circle route?
- How do percentages work? (Why does 5% added then taken away not give one back the original cost?)

Eventually, of course, higher Mathematics explains and extends the Mathematical ideas at lower levels, through proofs and investigations. As someone who taught *New Math* in the U.S. in the early 1960s, this idea of Mathematics as explainer resonates with the valuing at that time of building proofs into every level of the Mathematics curriculum. Certainly the role of proof marks out Mathematics as different from Science—science explains, but not by proof, more by experiment and data analysis. Mathematics explains both by being theory, and reflexively by proof.

Note, however, that I am not recommending a full-scale return to the challenges of trying to teach New Math to the whole school population. The relevant problem with the New Math approach was the scant reference to the real world, or to the scientific world. There was also no overt reference to the values which it was promoting. It was not a questioning curriculum.

Nor am I recommending a Back-to-Basics approach either. What I am doing, in the spirit of this book, is to support the core idea, and the various proposals in the book, for a more explicit consideration of the values and the philosophies behind mathematics education. This is not to satisfy a desire for mathematical navel gazing, but to explore how to enable mathematics education to revitalize itself and help tomorrow's students better understand why mathematical knowledge and values are so important in their world. Exploring the Mathematics/numeracy relationship is but one example of the necessary revitalization.

If, as I believe, values are the vehicles for approaching the questioning of philosophies of mathematics education, then there are many questions for our research agendas:

- What are these values, and where do they come from? I believe we have made a start with this question through historical analysis of mathematical writings, but much remains to be done.

- How are these values being "taught" (if at all)? Does "teaching values" mean anything sensible? Are values things which can be taught?
- How do we know if they have been learned? If we have a problem with whether values can be taught, do we nevertheless feel comfortable with the fact that values can be, and are, learned?
- If we accept that values can be learned, are we happy with which values are being learned?
- Are teachers responsible for this values teaching/learning?
- How much of values learning comes from "teachers" other than professional teachers, for example, parents, the media, communities?
- How much of this learning comes from other learners?

In the spirit of this book, I will leave these questions hanging, so to speak. Progress in knowledge is not just a matter of problem solving, or of finding new answers to old questions; it is mainly about posing the right problems and asking the right questions. Whether the questions asked here, or elsewhere in this book, are the right ones, is then the ultimate question for the reader to answer.

Name Index

Subject Index

Mathematics Education Library

Managing Editor: A.J. Bishop, Melbourne, Australia

---

1. H. Freudenthal: Didactical Phenomenology of Mathematical Structures. 1983 ISBN 90-277-1535-1 HB; 90-277-2261-7 PB
2. B. Christiansen, A. G. Howson and M. Otte (eds.): Perspectives on Mathematics Education. Papers submitted by Members of the Bacomet Group. 1986 ISBN 90-277-1929-2 HB; 90-277-2118-1 PB
3. A. Treffers: Three Dimensions. A Model of Goal and Theory Description in Mathematics Instruction TheWiskobas Project. 1987 ISBN 90-277-2165-3
4. S. Mellin-Olsen: The Politics of Mathematics Education. 1987 ISBN 90-277-2350-8
5. E. Fischbein: Intuition in Science and Mathematics. An Educational Approach. 1987 ISBN 90-277-2506-3
6. A.J. Bishop: Mathematical Enculturation. A Cultural Perspective on Mathematics Education.1988 ISBN 90-277-2646-9 HB; 1991 0-7923-1270-8 PB
7. E. von Glasersfeld (ed.): Radical Constructivism in Mathematics Education. 1991 ISBN 0-7923-1257-0
8. L. Streefland: Fractions in Realistic Mathematics Education. A Paradigm of Developmental Research. 1991 ISBN 0-7923-1282-1
9. H. Freudenthal: Revisiting Mathematics Education. China Lectures. 1991 ISBN 0-7923-1299-6
10. A.J. Bishop, S. Mellin-Olsen and J. van Dormolen (eds.): Mathematical Knowledge: Its Growth Through Teaching. 1991 ISBN 0-7923-1344-5
11. D. Tall (ed.): Advanced Mathematical Thinking. 1991 ISBN 0-7923-1456-5
12. R. Kapadia and M. Borovcnik (eds.): Chance Encounters: Probability in Education. 1991 ISBN 0-7923-1474-3
13. R. Biehler, R.W. Scholz, R. Sträßer and B. Winkelmann (eds.): Didactics of Mathematics as a Scientific Discipline. 1994 ISBN 0-7923-2613-X
14. S. Lerman (ed.): Cultural Perspectives on the Mathematics Classroom. 1994 ISBN 0-7923-2931-7
15. O. Skovsmose: Towards a Philosophy of Critical Mathematics Education. 1994 ISBN 0-7923-2932-5
16. H. Mansfield, N.A. Pateman and N. Bednarz (eds.): Mathematics for Tomorrow's Young Children. International Perspectives on Curriculum. 1996 ISBN 0-7923-3998-3
17. R. Noss and C. Hoyles:Windows on Mathematical Meanings. Learning Cultures and Computers. 1996 ISBN 0-7923-4073-6 HB; 0-7923-4074-4 PB
18. N. Bednarz, C. Kieran and L. Lee (eds.): Approaches to Algebra. Perspectives for Research and Teaching.1996 ISBN 0-7923-4145-7 HB; 0-7923-4168-6, PB
19. G. Brousseau: Theory of Didactical Situations in Mathematics. Didactique des Mathématiques 1970-1990. Edited and translated by N. Balacheff, M. Cooper, R. Sutherland and V. Warfield. 1997 ISBN 0-7923-4526-6
20. T. Brown: Mathematics Education and Language. Interpreting Hermeneutics and Post-Structuralism. 1997 ISBN 0-7923-4554-1 HB
    Second Revised Edition. 2001 ISBN 0-7923-6969-6 PB

Mathematics Education Library

21. D. Coben, J. O'Donoghue and G.E. FitzSimons (eds.): Perspectives on Adults Learning Mathematics. Research and Practice. 2000 ISBN 0-7923-6415-5
22. R. Sutherland, T. Rojano, A. Bell and R. Lins (eds.): Perspectives on School Algebra. 2000 ISBN 0-7923-6462-7
23. J.-L. Dorier (ed.): On the Teaching of Linear Algebra. 2000 ISBN 0-7923-6539-9
24. A. Bessot and J. Ridgway (eds.): Education for Mathematics in the Workplace. 2000 ISBN 0-7923-6663-8
25. D. Clarke (ed.): Perspectives on Practice and Meaning in Mathematics and Science Classrooms. 2001 ISBN 0-7923-6938-6 HB; 0-7923-6939-4 PB
26. J. Adler: Teaching Mathematics in Multilingual Classrooms. 2001 ISBN 0-7923-7079-1 HB; 0-7923-7080-5 PB
27. G. de Abreu, A.J. Bishop and N.C. Presmeg (eds.): Transitions Between Contexts of Mathematical Practices. 2001 ISBN 0-7923-7185-2
28. G.E. FitzSimons:What Counts as Mathematics? Technologies of Power in Adult and Vocational Education. 2002 ISBN 1-4020-0668-3
29. H.Alrø andO. Skovsmose:Dialogue and Learning in Mathematics Education. Intention, Reflection,Critique.2002 ISBN 1-4020-0998-4 HB;1-4020-1927-0 PB
30. K. Gravemeijer, R. Lehrer, B. van Oers and L. Verschaffel (eds.): Symbolizing, Modeling and Tool Use in Mathematics Education. 2002 ISBN 1-4020-1032-X
31. G.C. Leder, E. Pehkonen and G. Tőrner (eds.): Beliefs: A Hidden Variable in Mathematics Education? 2002 ISBN 1-4020-1057-5 HB; 1-4020-1058-3 PB
32. R. Vithal: In Search of a Pedagogy of Conflict and Dialogue for Mathematics Education. 2003 ISBN 1-4020-1504-6
33. H.W. Heymann: Why Teach Mathematics? A Focus on General Education. 2003 Translated by Thomas LaPresti ISBN 1-4020-1786-3
34. L. Burton: Mathematicians as Enquirers: Learning about Learning Mathematics. 2004 ISBN 1-4020-7853-6 HB; 1-4020-7859-5 PB
35. P. Valero, R. Zevenbergen (eds.): Researching the Socio-Political Dimensions of Mathematics Education: Issues of Power in Theory and Methodology. 2004 ISBN 1-4020-7906-0
36. D. Guin, K. Ruthven, L. Trouche (eds.) The Didactical Challenge of Symbolic Calculators: Turning a Computational Device into a Mathematical Instrument 2005 ISBN 0-387-23158-7
37. J. Kilpatrick, C. Hoyles, O. Skovsmose (eds. in collaboration with Paola Valero) Meaning in Mathematics Education. 2005 ISBN 0-387-24039-X
38. H. Steinbring: The Construction of New Mathematical Knowledge in Classroom Interaction: An Epistemological Perspective. 2005. ISBN 0-387-24251-1
39. M.Borba, M. Villarreal: Humans-with-Media and the Reorganization of Mathematical Thinking: Information and Communication Technologies, Modeling, Visualization and Experimentation. 2005 ISBN 0-387-24263-5 HB, ISBN 0-387-32821-1 PB
40. G. Jones (ed): Exploring Probability in School: Challenges for Teaching and Learning. 2005 ISBN 0-387-24529-4
41. D. DeBock, W. Van Dooren, D. Janssens, and L. Verschaffel: The Illusion of Linearity: From Analysis to Improvement. 2007 ISBN 978-0-387-71082-2
42. K. François and J. P. Van Bendegem: Philosophical Dimensions in Mathematics Education. 2007 ISBN 978-0-387-71571-1